# FOOD, GENES, AND CULTURE

# FOOD, GENES, AND CULTURE

## EATING RIGHT FOR YOUR ORIGINS

*Gary Paul Nabhan*

ISLANDPRESS Washington | Covelo | London

Library of Congress Cataloging-in-Publication Data

Nabhan, Gary Paul.
Food, genes, and culture : eating right for your origins / Gary Paul Nabhan.
pages cm
Includes bibliographical references and index.
ISBN 978-1-61091-492-5 (pbk.) -- ISBN 1-61091-492-9 (paper) 1. Nutrition--Genetic aspects. 2. Food habits--History. 3. Human evolution. 4. Physical anthropology. I. Title.
QP144.G45N33 2013
612.3--dc23

2013023650

British Cataloguing-in-Publication data available.

Printed on recycled, acid-free paper

Design by Kathleen Szawiola

Manufactured in the United States of America
10 9 8 7 6 5 4 3 2 1

*"Now killin' folks and cookin'*
*ain't so very far apart."*

HERBERT H. KNIBBS

*from his poem/song "Boomer Johnson"*

*Dedicated to the memory of*
*Ted Nabhan,*
*Sally Giff Pablo, and*
*Gabriel Williams*

CONTENTS

# FOREWORD

IF THERE IS A MORAL to this story, it is that we ignore the interactions among our genes, our ancestral and contemporary diets, and our environments (including their myriad microbes) at our own peril. But if there is hope in this same story, it is that once we open our eyes, mouths, and taste buds to these fascinating interactions, our world will be made richer and many problems can be averted.

What kinds of problems? For starters, the human suffering triggered by the onset of diabetes, heart disease, food allergies, and many forms of diet-driven inflammation. Over the long haul, we also need to check the decline of biodiversity which will impoverish us all, but particularly that of the place-based microbes in our food systems, from the bacteria in our garden's soils to those in our guts.

These diseases and degradations affect the quality of life of billions of people, yet they are often mislabeled if not misdiagnosed and attributed to the wrong causes. Take adult-onset or non-insulin dependent diabetes, for example, which medical researchers have treated either as a genetically determined disease or a nutritional "disease of Western civilization," rather than as the result of mismatches between genes, environment, and diet. Since the first edition of this book was published, the number of Americans living with diabetes has grown to an estimated 22.3 million—about 7% of the U.S. population. That marks an increase of nearly five million or 22 percent from 2007 to 2012 and, according to the American Diabetes Association, translates to $245 billion of medical costs. If no change in our perspective on this disease occurs by 2030, Americans might pay as much as $1.3 trillion annually for its treatment. Already, the collateral damage generated by our unhealthy food system means that one in every four dollars spent on hospital care in America is spent on diabetes, and one in every ten dollars spent on health care in general is spent on this same disease of gene-food-environment dysfunctions. What if just $50 million a year were dedicated to redesigning the food production, processing, and preparation system that is making our citizens sick in the first place?

But just why have we witnessed such astronomical growth in diabetes and other diseases during our own lifetimes? While many have suspected that there is a tangible link between this disease and nutrition, we have not yet dealt with its root causes. As a result, 95% percent of U.S. cases of diabetes are clearly triggered by diet changes that result in the non-insulin-dependent diabetes mellitus (NIDDM) form of this disease. Most experts point to the rise in consumption of high-fructose sugars, but what's painfully absent from the discussion is the connection between health and food, mediated by our genes, culture, and environment. Most geneticists once understood that any expression of certain diseases in plants or animals resulted from at least three factors: genes, environment, and gene-environment interactions. Today we can also include the epigenetic consequences of microbial (biotic) and climatic (abiotic) influences.

Where I live in Arizona, among many diverse Native and Hispanic American populations that suffer the highest incidences of diabetes anywhere in the country, we are headed toward a *nutritional cliff* that is also a "true" *fiscal cliff* . Diabetes care in my state has jumped from $.5 billion in 1995 to $3 billion in 2005, to at least $4.4 billion by the end of 2013. To put that in perspective, for every $20 dollars of food grown in Arizona, at least one dollar goes to treating diabetes sufferers.

If diabetes were the only problem, perhaps such bad news could be swept under the rug. But instead, Americans are now suffering from what Moises Velasquez-Manoff has called *an epidemic of absence* that can explain most of the meteoric rises not only in adult-onset diabetes, but in allergies, auto-immune diseases, and inflammatory diseases such as colitis that have occurred over the last half century. Yet the well-documented interactions among human genes, gut microbes, place-based foods, and diseases have barely changed the way most Americans think about our food choices. Sadly, even influential American food writers such as Michael Pollan and Marian Nestle fail to feature topics such as dietary diversity and human genetic diversity in their prescriptive "food rules;" for them, it is enough merely to get Americans to count "good" and "bad" calories, let alone worry about the food-microbe-gene interactions which affect how most of us respond to those calories.

Societies outside North America and Western Europe appear more cognizant of the connections between food, cultural heritage, and place than we are. The release of the first edition of this book stimulated far more public discussion abroad than in the U.S. In

Italy, it was discussed in more than fifty papers, on several television stations, and in Slow Food circles. In Mexico, the national Fondo de Cultura Económica selected it in 2006 as *the* book on science and society from elsewhere in the world to be translated into Spanish. The difference in reception says less about the book itself than about its readers: in Italy and Mexico, most residents both emotionally identify with and intellectually understand deeply historic food traditions. In much of the United States, our citizenry has become so placeless that I once heard a young Slow Food USA employee ask whether we even had "heritage foods" linked to our own landscapes and communities in North America! And since in America, many consider ourselves "mutts" or genetic hybrids of mixed ancestry, few fathom that our own unique sets of genes may be interacting with particular foods (or the lack of them) in ways which profoundly influence health.

Yet important research is being done both inside the U.S. and abroad in the fields now known as *nutritional epigenetics, evolutionary gastronomy,* and *ecological genetics.* Tufts University researchers Sang-Woon Choi and Simonetta Friso have called epigentics, in particular, "the new bridge between nutrition and health." Epigenetics has been described as the heritability of gene expression that can result from environmental or nutritional influences that do not actually alter DNA sequences or genomic structures. In improving nutrition, Choi and Friso contend that epigenetics can be exceptionally useful, because the right nutrients and microbially-enhanced probiotic foods can dramatically influence epigenetic phenomena and trigger the expression of genes at the transcriptional level. In essence, they are suggesting that diets consistent with our ancestry can positively reset genetic expression as a means for restoring our health.

The idea that eating nutritionally and culturally-appropriate foods can reset our genes toward wellness is being aggressively explored by a growing number of nutritional epigeneticists all around the world. Since the publication of the first edition, numerous community health programs among Hispanic and Native Americans are "getting back to their food roots," as Rebecca Wiggins-Reinhard of Somos La Semilla Food Center calls these grassroots but culturally- and scientifically-informed initiatives. Rebecca and I have been particularly impressed and inspired by the efforts of Rubi Orosco, a public health specialist with the non-profit Mujer Obrera in El Paso Texas. As Rubi once explained,

> "When people think about eating healthy, they often think about having to eat things that are very foreign to the way they eat now, like wheat grass or soy. In reality, we just have to look back a couple of generations, to go back to what our grandparents were doing. The food is still familiar enough, and is in our genes…"

Both conventional medical doctors and community health practices are becoming more open to the value of an evolutionary and revolutionary gastronomic approach that finds a new balance between traditional cultural knowledge and cutting-edge science. Interventions such as the Perfect Health Diet of Paul and Shou-Ching Shih

Jaminet and the BalancePoint diet of biochemist Binx Selby are but two of many empirical demonstrations that we can reverse diseases such as diabetes and antherosclerosis through cuisines in harmony with our genes. They offer an approach that seeks to honor the time-tried relationships between the diverse edible floras, faunas, and microbial communities in our regional foodscapes and in our guts as well. But it is also an approach which humbly admits that we must seek to understand rather than ignore the relationships between bio-complexity, our own health, and that of generations to come.

GARY PAUL NABHAN
Kellogg Endowed Chair in
Food and Water Security
for the Borderlands
UNIVERSITY OF ARIZONA

# INTRODUCTION

I AM ABOUT to take you with me on a culinary and evolutionary odyssey, one that will reveal that our ancestral homelands do not lie in some remote, nearly unreachable place, but instead are imbedded in our genes and our cultural food preferences. That is to say, there are dynamic connections between our culinary predilections, our genes, the diets of our ancestors, and the places that our ancestral cultures called home for extended periods of time.

As we set out to fathom the depth of these connections, we will visit island after island and continent after continent where parts of the story are most vividly apparent. At each of these localities, we will come to see that each distinctive ethnic food tradition around the world does not simply consist of random ingredients brought together through some serendipitous experimentation by a master chef. Instead, each ethnic cuisine reflects the evolutionary history of a particular human population as it responded to the availability of edible plants and animals through local foraging and through trade, and to the prevailing frequencies of diseases, droughts, and plagues within each population's homeland. Our odyssey will reveal the tragedies these ethnic populations have suffered whenever they have been displaced from their homelands or have strayed too far from their traditions, seduced by foreign food and drink. But our journey will also celebrate a kind of *homecoming*—times when we may feel that our genes, our cultural traditions, and the foods we eat all come into perfect alignment with one another—so that the health of our bodies, communities, and habitats are one and the same.

I have viscerally sensed this sort of homecoming when I have eaten with my Lebanese relatives foods that have been cultivated and cured in the Fertile Crescent for thousands of years; I have also seen Native Americans celebrate such homecomings when they have prepared village feasts comprised of wild foods that were pit-roasted around camps in desert homelands long before the first Europeans and Moors arrived. Perhaps these intuitions and experiences have been my greatest motivation for setting this story down; I want to understand the source of the "extensive" pleasure I have felt when participating in such feasts. Ever since I was a child growing up in an urban melting pot, I have been intrigued by ethnic food traditions. I remember how different the holiday foods of my neighbors were from one another, as I moved between homes of immigrant Greeks, Polish Jews, Sicilians, the Irish, Swedes, and Mexicans. Later, as I grounded myself in genetics, ecology, anthropology, and nutritional sciences, I realized that there was a unifying theme linking much of my scholarship and field research: how food reflects the interaction between biological and cultural diversity.

There was a pivotal moment during my graduate school days that forced me to see the parallels and differences among various ethnic food traditions. It occurred when a friend and I were invited for Thanksgiving dinner by a Native American family, a clan of Pima Indians that lived out in the desert south of Phoenix, Arizona. As the eldest woman in the clan prepared tepary beans, mesquite pudding, and other traditional foods that had been eaten in the region for millennia, my friend Amadeo mentioned to her that I was Lebanese.

"Oh, that Lebanese cooking," she said, "it has such interesting spices!" I was taken by surprise that she even knew what Lebanese cooking was, until her daughter explained to me that she had cooked for a Lebanese lawyer in Phoenix for a number of years. Then the elderly lady turned away from the stove and—brandishing ladle in hand—looked me in the eye and asked, "Do you grow those spices or get them from your relatives? I mean, they must be hard to get here in Arizona. . . . It must be like our own native foods that will disappear if no one keeps the tradition going. . . . That's when our health goes down the drain even more."

At that moment, I realized that most truly authentic ethnic cuisines are in peril, even though they have displayed resiliency and continuity over hundreds if not thousands of years. Furthermore, a certain cultural loss occurs when such foods are abandoned by a particular community, whether through assimilation or as a result of refugees fleeing their homeland. If we are to prevent such losses, we must more deeply understand why ethnic food traditions *matter*.

And so, I set off to prepare for this present journey. Like any good odyssey, this story twists around on itself as if it were the double helix itself, the DNA that snakes through your body and mine like a meandering river of memories about our ancestors, the places they lived, what they ate, and at times, about what they had to reject or eject from their bodies. In our lifetimes, we loop back to taste certain delicacies more than once, but we also recoil from culinary perils along the way. That is to say, *we are what our ancestors ate, and also what they had to regurgitate*, for there are as many poisonous plants, fungi, fish, and shellfish surrounding us as there are edible, delectable ones. Our ancestors developed their own traditions of ecological knowledge to discern the delicious and nutritious from the toxic. This knowledge not only helped them select edible foods from the bewildering diversity of flora and fauna within their reach; as we shall see, it also served to foster cultural diversity, as distinctive food foraging and cultivation traditions emerged that set various ethnic populations on very different evolutionary trajectories. It may well be that certain secondary compounds in staple foods and culinary herbs induced mutations among certain populations, and that some of these genetic changes resulted in differential survival and reproductive success among a population's individuals. In short, natural selection and other evolutionary processes mediated by food choices have likely played important roles in generating both human genetic diversity and orally transmitted cultural diversity.

This journey will force us to confront dangers far more insidious than poisonous plants and animals. We will also have to face the fatalism engendered by genetic determinists and the racism promulgated by the advocates of eugenics. In addition, we must face our own gullibility; that is, how easily many of us have been seduced by one-size-fits-all fad diets or by promises of quick genetic fixes.

As we launch this odyssey together, we will take with us a new tool to help us navigate past these pitfalls. Like many tools, this one is a double-edged sword—it has the capacity to help us heal, but if used improperly, it can kill. The new tool is a database nicknamed OMIM, short for Online Mendelian Inheritance in Man, the internet-accessible human genome database prepared by the National Center for Biotechnology. Edited by Victor McKusick at Johns Hopkins, it includes a list of "disease genes" organized so that you may search for the gory details of any heritable human malady that captures your attention or concern. The Web site also includes the OMIM Gene Map, a chromosome-by-chromosome location of all genetic disorders recently described and compiled by the Human Genome Project.

These so-called maps did not initially make much sense to me, perhaps because I am more intrinsically interested in how these genes are distributed across the face of the Earth than I am in their distribution across the face of my very own chromosomes. But when I started scrolling through the list of various disease genes, I quickly became intrigued by the oddity of language that human geneticists had used in describing their discoveries.

The names of the genes were deliciously stuffed with irony. There was the "Maple Syrup Urine Disease" gene and the "Café-au-Lait Spots, Multiple with Leukemia." Then there was the gene called "Acromesomelic Dyplasia, Hunter-Thompson type," discovered in the days when an aspiring young journalist named Hunter Thompson raised funds to support his writing habit—among other habits—by giving blood for use in genetic research. There was also a gene for the

"Novelty Seeking Personality Trait," one that I'm sure Hunter Thompson *wished* had been named for him.

But what riveted my attention was a particular subset of "genetic disorders"—ones that interact with the foods we eat and the beverages we drink in ways that either make us sick or keep us well. That list, placed at the end of this introduction (table 1), suggested that certain foods must have interacted with particular sets of genes *over the course of human evolution*. Nevertheless, as I compiled this list from the OMIM Web site, I was struck by various commentaries imbedded in the texts describing each gene. I realized that in the heady euphoria of biological discovery, a number of geneticists believed that they had found genes that *caused* alcoholism, as well as one that conferred alcohol intolerance. However, as I read more of the texts about these genes, their stories seemed somewhat fuzzier than anything as simple as a one gene/one disease relationship. Several genes and many cultural variables seemed to be involved in getting us drunk or keeping us sober.

I was also surprised to see that there is a gene for sucrose intolerance; how, I wondered, did that gene survive the twentieth century, when just about everything from toothpaste to postage-stamp glue came to have sugar in them? Another gene on the list makes many people intolerant of the gluten in cereal grains, which contributes to celiac disease. But as I read deeper into the OMIM texts, I found that like alcoholism, celiac disease is not strictly determined by genetic factors alone. As Victor McKusick cautioned his readers, "celiac disease, also known as celiac sprue and gluten-sensitive enteropathy, is a multifactorial disorder of the small intestine that is influenced by both environmental and genetic factors."

In *The Dependent Gene*, David Moore has underscored how critical it is that the public understand that simple phrase, *influenced by both environmental and genetic factors*: "Nearly every day at the beginning of this millennium, we are encountering news reports of the discovery of the gene 'for' some human trait or illness. . . . Unfortunately, these astonishing advances have been presented to the public in a way that has perpetuated this mistaken idea that some of our traits are caused exclusively (or primarily) by our genes. In fact . . . all of our traits—bar none—emerge from the mutually-dependent activity of both genetic and environmental factors" (Moore 2001).

From the OMIM Web site, I gleaned that there are no less than twenty-six genes on sixteen chromosomes that interact with various environmental factors, namely with the foods and beverages characteristic of certain ethnic diets rooted in particular places around the world. Many of these are polymorphic, taking various forms, each of which interacts with dietary chemicals in slightly different ways. In some cases, the combination of a particular gene and the presence of a particular food in an ethnic diet can protect many individuals of that ethnicity from an infectious or nutritional disease. In other cases, a gene-food interaction can literally kill the carrier of a particular allele, or gene variant. At the same time, those carrying other alleles suffer no health risks at all. There are dozens of variations on these themes, depending upon whether one or more genes are involved as well as the potency of the food, beverage, or drug that the carrier ingests.

Having taken several genetics courses in college and coauthored several scientific articles on wild-plant genetics, this did not surprise me. I knew intellectually that certain genes encode the ways in which our bodies produce the enzymes that drive our metabolisms. And I emotionally accepted that the genetic lot that each of us is cast may mean that some of us produce a paucity of a certain enzyme required for necessary bodily functions, while others produce the same enzyme in abundance. For instance, I must concede that as someone who suffers from attention-deficit alternating with hyper-focus, my dopamine levels are often wildly variable compared to my even-keeled colleagues, just as I accept that my given lot includes red-green color- blindness. When such variations are identified, medical researchers often conclude that the enzyme-deficient carrier has a "genetic disorder"—one that can make him or her susceptible to a number of perils: nutritional imbalances, mood swings, or even heart disease and cancer.

On the other hand, is it a *disorder* if the lack of production of an enzyme protects some of us from a disease such as malaria? One enzyme deficiency known as favism does just this, reducing its carriers' vulnerability to malaria, the number one killer in the Mediterranean basin over the last ten millennia. For that matter, is it accurate to call some condition a *genetic* disorder if the functioning of a gene only kicks in when particular ethnic foods are regularly consumed in traditional diets?

What I am getting at is this: it seems there are some conditions that scientists once simplistically lumped together as *genetic disorders* that instead might be considered *environmentally specific adaptations* that actually increase our fitness in certain settings or on certain diets. I am not the only biologist skeptical of such cut-and-dried categories. Today, there are a growing number of scientists from many disciplines that consider gene-food interactions to be *adaptations* in certain contexts, and *disorders* in others. Up until recently, these scientists called their field *nutritional ecogenetics*; today, those on the high-tech side of things call their suddenly lucrative business *nutrigenomics*.

As you are probably aware, humans living on this planet do not share all of the same genes, nor do they necessarily favor the same foods. While about 85 percent of human genetic diversity can be represented by different individuals in the same ethnic population, as much as 15 percent appears as differences among or between various ethnic populations. This is, in short, evidence of people's divergence from their common ancestors as they came to live in different places, where they came to be exposed to different food choices and diseases. While there are many ways of defining ethnicity, one way is by language, and some 6,500 different languages remain spoken on this planet by humans (although that number may be halved by the end of this century). Each of these language groups has a different way of speaking about food, of collecting and preparing it, and a different vocabulary to describe its cultural identity in relation to the foods it favors.

And so, this meandering story is one about human genetic diversity interacting with the diverse cultural traditions of food getting, food preparation, and food consumption. While some genes and ethnic cuisines may persist because they clearly have had adaptive value in certain settings, many scientists doubt that the evolutionary process of adaptation is the only explanation for the set of traits we carry or the set of foods we most cherish on the dinner table. It could very well be that certain genes emerged through random "mutations" and are of neutral or negligible value with respect to our survival. By analogy, we might say that some features of ethnic cuisines persist because they aided our survival under particular environmental stresses, while others are more "ornamental," like the proverbial icing on the cake. Nevertheless, more and more scientists now accept that ethnic cuisines have deep-seated ecological underpinnings and evolutionary trajectories that are of great significance to the health status of their consumers. The scientists who muse over such issues have recently begun to call themselves Darwinian gastronomists, but as I explain later in this book, a broader term—evolutionary gastronomists—is perhaps more apt, for it acknowledges that some gene-food interactions emerged by evolutionary processes unrecognized at the time of Darwin, processes that have led to discernible genetic changes in ethnic populations in 1,500 years or less.

If you are wondering what this has to do with *you*, the answer is simple: it depends on just which genes you carry. Perhaps the baldest way to assess the importance of gene-food-culture interactions is to consider how many people are prone to sickness or death when their cuisines and cultures get out of sync with their genes. Today, more than 100 million people suffer from adult-onset diabetes, a nutritional disease to which some ethnic populations are genetically predisposed more than others. Lactose intolerance affects upwards of 2.25 billion adults and 600 million others under the age of twenty—roughly one-half of the world's population. Heritable food allergies under genetic influence affect no less than 200 million individuals worldwide. At least 100 million people have a deficiency of an enzyme called G6PD, which interacts with fava beans, culinary and medicinal herbs, as well as over-the-counter medicines. Perhaps 50 million Europeans and European Americans have unusually high levels of homocystine and heart disease unless they regularly ingest greens and beans rich in folic acid. Throw in alcohol intolerance, gluten intolerance, fructose and maltose intolerance—or a host of other genetic interactions with particular foods, beverages, or the chemical compounds contained within them—and well over three-quarters of the world's population carries one or more of these so-called genetic disorders. It may therefore be reassuring for you to know that if you have such a "disorder," statistically speaking, you are a "normal" human being.

To understand just how profoundly gene-food-culture interactions affect humankind, we must venture forth and listen to the stories of people who suffer from such interactions or are protected by them. We must grapple with the history of these genes, ensuring that it is a contextual history with a human face and a memorable natural setting. The places in many of the chapters that follow happen to be islands: Java, Bali, Crete, Sardinia, and Hawaii. That is because many traits become fixed more rapidly in a larger portion of a population that is largely restricted to one or more islands than they do on large continents where genetic intermixing is more pervasive. The peculiar evolutionary histories of plants and animals on islands often means that they are chock-full of peculiar chemicals that make their uses as human food and medicine even more interesting and exotic.

As I make final preparations for our journey together, I find myself amazed by how many genes have set groups of people apart from others and by how weirdly these genes have responded to certain dominant foods in particular ethnic diets. If I tried to make up stories as odd as some of the ones you are about to read, I doubt that anyone would believe me. They remind us of just how diverse humankind is in its genes, its tastes, and its ethnic histories. This journey into our genetic and culinary histories celebrates these differences and, I hope, effectively cautions against thinking that quick genetic fixes will stave off human suffering without leading to other kinds of problems for us and for future generations.

TABLE 1 • A REGISTRY OF GENE-FOOD INTERACTIONS ASSOCIATED WITH THE HUMAN FAMILY TREE

| DISORDER/ADAPTATION | GENE MAP LOCUS | DEMOGRAPHIC DATA | FOOD/DRUG/ BEVERAGE TRIGGER |
|---|---|---|---|
| Alcoholism | Many, incl. chrs. 4p, 4q22, 17q21, 11q23, 11p15, 22q11 | Broad; (Native) America, Asia, Australia | Fermented grains and tubers |
| Alcohol dehydrogenase (ADH2) | Chr. 4q22, 11s | Broad; (Native) America, Asia, Australia | Fermented grains and tubers |
| Aldehyde dehydrogenase variant (ALDH1Aa) | Chr. 9q21 | Broad; (Native) America, Asia, Australia | Fermented grains and tubers |
| Aldehyde dehydrogenase variant (ALDH2) | Chr. 12q24 | Japan, China, South America | Fermented grains and tubers |
| Amotrophic lateral sclerosis–Parkinsonia-dementia (ALS–PD) | Chr. 17q21.1 | Guam, Kii peninsula of Japan | Cycad seeds, flying foxes, which ingest cycad seeds |
| Apolipoprotein A | Chr. 11q23 | Europe and elsewhere | Vegetable and animal fats |
| Apolipoprotein B | Chr. 2p24 | Europe and elsewhere | Vegetable and animal fats |
| Apolipoprotein E (APOE2) | Chr. 19q13 | Broad; esp. Mediterranean | Vegetable and animal fats |
| Celiac disease (gluten sensitivity) | Chr. 6p21 | Europe, North America | Gluten from wheat, rye, barley |
| Cytochrome P450 (coumarin 7-hydroxylase) | Chr. 19q13 | Many variants; Central Asia, China | Coumarin in herbs, key vegetables, fruits |
| Diabetes mellitus type 2 (NIDDM) | Many, incl. chrs. 2q32, 11q12, 13q24, 17q25, 20q | Broad, many variants; (Native) America, Australia, Polynesia | Fast-release, fiber-depleted foods |
| Disaccharide intolerance (sucrose-isomaltose malabsorption) | Chrs. 3q22–q26 | Native America, incl. Inuit (Eskimo), Greenland, Siberia | Milk, sucrose, maltose in high concentrations or quantities |
| Fanconi-Bikel syndrome | Chr. 3q26 | Scattered; Swiss Alps, Japan | Galactose sugars |
| Fructose intolerance | Chr. 9q | British Isles | Fruits |
| Glucose-6-phosphate dehydrogenase (G6PD, favism) | Chr. Xq28 | Mediterranean | Fava beans, antimalarial drugs, some herbs |

CONTINUED

TABLE 1 • CONTINUED

| DISORDER/ADAPTATION | GENE MAP LOCUS | DEMOGRAPHIC DATA | FOOD/DRUG/ BEVERAGE TRIGGER |
|---|---|---|---|
| Homocysteinemia | Several, incl. chr. 21q22 | Broad: Europe, America | Vitamin B12 |
| Homocystinura | Chrs. 21q22, 5p15 | Northern Europe, British Isles | Lack of folic acid from greens, beans |
| Hypercholes-terolemia | Chr. 2p24 | Northern Europe, British Isles, America | Fast-release, fiber-depleted foods |
| Insulin resistance | Chr. 11p15 | Broad, many variants; (Native) America, Australia, Africa | Fast-release, fiber-depleted foods |
| Lactose intolerance | Chr. 2q21 | Lactase persistence in Northern Europe, Arabia, parts of Africa; deficiency everywhere else | Milk products |
| Phenylthiocarbamide tasting (PTC tasting/ PROP tasting) | Chr. 5p15 | Broad, many variants | Chiles, quinine, certain drugs and bitter herbs |
| Serum albumin A | Chr. 4q11, chr. 7 | Many variants; Eti Turks and others in central Asia, Athapaskan and Uto-Aztecan in the Americas | Coumarin-containing plants and drugs, incl. sage, warfarin |
| Transferrin | Chr. 3q21 | Africa, esp. Zimbabwe | Diets deficient in vitamin C and iron |

SOURCE: V. McKusick, Online Mendelian Inheritance in Man database, URL.

LEGEND: *Column 1* lists the common name for a particular medical condition, with its more precise technical or other abbreviation in parentheses; although most physicians consider these conditions to be disorders, some may be place- or diet-specific adaptations to diseases or stresses.

*Column 2* cites the gene map loci that confer the condition, noting both the chromosome(s)—chr. or chrs.—and the general or specific location of the gene(s) presumed to be involved. For instance, Aldehyde dehydrogenase variant 1Aa is located on chromosome 9 at the q21 locus. See the OMIM database for more detail, as well as for the published sources that first related these loci to particular medical conditions.

*Column 3* indicates, where possible, the human populations with particular geographic distributions that tend to have higher frequencies of individuals carrying the gene(s) or any variants. (Native) America refers to American Indian, Inuit, and related populations, where America refers to a condition also shared with Euro-, African-, and Asian-Americans.

*Column 4* lists some but not all of the foods and drugs (or deficiencies of same) that interact with the gene(s) in ways that alter the health status of the carrier individual. Note that in nearly all cases, a single gene in and of itself does not, for example, "cause" alcoholism. Most of these conditions are influenced by multiple genes and by environmental, cultural, and developmental conditions.

CHAPTER ONE

# *Discerning the Histories Encoded in Our Bodies*

AS WE INITIATE our journey together, I want to take you to a place in the desert that forced me to embark on this journey to begin with. It was there that I first saw for myself how the inexorable loss of ethnic food traditions could send, as Pima Indian elder had prophesied to me, the health of an entire people down the drain.

It was another Pima Indian friend, a fellow gardener named Gabriel, who first made me see that various ethnicities respond in dramatically different ways to the very same foods and drinks. Only later did I understand the degree to which these responses are curious outcomes of interactions among genes, environments, and cultures, some of them tragic, some protective, and others downright funny. While Gabriel was the first friend I lost to the darker side of these interactions, he was also the first to let me see the lighter side. He did so in a way that was patterned after the behavior of that old-time trickster, Coyote.

"Hey, White Guy, can you take some time out from your busy schedule to help me drive some commodity foods out to some bro's of mine out at Ak-Chiñ village?"

"Sure, I'll be your delivery boy. What are we going to deliver? Italian pizza or Indian fry bread?"

"Powdered milk. A bunch of it. Help me put it in the back of your pickup."

The powdered milk came from the federal government surplus commodity foods program, which typically provided such foods to low-income families on the reservation on a monthly basis. The foods were also stockpiled in a warehouse where families had to come with their vouchers to obtain them, but because Gabriel worked for the tribe's nutrition program, he had a little of every commodity stashed away in a storage closet in his office. On occasion, we would share the hidden stash with his friends out in remote villages who did not come into town very often; sometimes we would even sneak the cans and boxes of government commodities into Mexico for Indian friends living south of the border. I did not particularly like the cans of greasy beef, the white flour, and the Velveeta-like cheese the government offered, not merely because none of these foods were part of the traditional Indian diet, but because many were fatty, sugary, or fiber depleted; in short, the kiss of death for Native American communities already suffering from nutrition-related diseases. In this case, powdered milk seemed like the least of the evils the commodities program had to offer, so I reluctantly helped Gabriel place several big cardboard boxes of white powder in the pickup truck. Then we were off;

driving down winding dirt roads through fields of desert wildflowers on a lovely spring morning.

As we arrived at Ak-Chiñ village, Gabriel directed me over to a baseball field where a number of young men and teenage boys appeared to be practicing for an upcoming game against another village's team. He got out, walked over and talked to one of the men in his native tongue, then came back to the truck.

"C'mon, White Guy, this is where we can leave the boxes. Gonna help me?"

I didn't get it. "You mean we're gonna distribute the milk here so that these guys can take it home to their families? Why don't we just drive around and drop the boxes off at their homes?"

Gabriel laughed wildly. "Noooo. They've had more of this crap at home than they know what to do with. It just sits there and goes bad. They stopped picking it up at the warehouse, but now they need some for the baseball game tonight."

"They're serving milk at a baseball game?"

"No, White Guy. They need it to lay down the baselines so the players will know where the infield is among all them wildflowers! Serve the stuff? You're kiddin', 'enit? We can't drink milk, even when it's mixed up from powder! Give me milk, and I bloat up like the Pillsbury Doughboy. Don't you know squat about us? All of us Indians got lactose intolerance."

Ah, lactose intolerance. Sure, I had heard that some Indians suffered from it, but could not recall any of them talking about it in front of me. It was years later before I understood that lactose intolerance is not just a dietary constraint for Pima Indians; more than 30 million Americans—including many of recent African and Asian descent—cannot digest the principal sugar in milk very long after they have been weaned from their own mothers' breasts. In fact, the weaning of most breast-fed children in the world may be precipitated by a gradual decline in the activity of lactase, an enzyme that breaks down the lactose into easily digestible glucose and galactose. Without sufficient lactase to digest it, lactose simply sits in the child's gut, absorbing water through osmosis and expanding until it forms a substrate for gas-producing microbes.

I realized that I was perhaps in the minority of Arizona residents whose tolerance to lactose extended into adulthood. Among Gabriel's Pima and Papago (O'odham) Indian kin, such lactose malabsorption affects 40 percent of all four-year-olds, 71 percent of all five-year-olds, 92 percent of all seven-year-olds, and 100 percent of the population eight years or older. If exposed to as little as four ounces of raw milk, both weaned children and adults suffer bloating, indigestion, and in severe cases, intestinal cramping and diarrhea.

Three decades ago, a cultural geographer named Frederick Simoons noticed that the global distribution of extended lactose tolerance was strongly correlated with the distribution of ancient herding peoples in Europe, Asia Minor, and northern Africa; teenage and adult residents in most of the rest of the world were lactase deficient. Around 10,000 years ago, a mutation occurred in the DNA of an isolated population of northern Europeans that allowed them to tolerate milk as a nutrient-rich resource. This adapted tolerance to milk gradually spread through intermarriage with other groups but may have independently emerged as a mutation in the DNA of other peoples as well. In any case, certain ethnic populations that carried this gene in low frequencies—and then subsequently adopted a pastoral lifestyle and cultured-milk consumption—found that their lactase activity gradually extended into adulthood. It is assumed that most of these people first used small quantities of raw milk in a ritualistic manner or initially consumed only fermented products such as yogurt and cheese, for which bacteria have already converted lactose into digestible sugars. The small percentage of lactose-tolerant individuals in any population was rapidly favored when these rich nutritional resources arrived, so that within just fifteen generations of eating cheese and yogurt, the frequency of lactose tolerance increased dramatically.

It appears that two single-unit DNA changes occurred that extended lactase enzyme production into adulthood among herding peoples. From an evolutionary perspective, it seems that lactose intolerance—which formerly regulated the time of weaning among nonagricultural societies—was suddenly relaxed. Keep in mind that among hunter-gatherers who had never kept livestock, children were typically weaned earlier than they were among herding societies. In wildland habitats where the supply of foods was seasonally variable, early onset of lactose intolerance would curb the child's desire to nurse and might keep mothers from depleting their reserves. This would also allow maternal fertility to resume earlier, since it is otherwise depressed by lactation. In short, childbirths in hunter-gatherer families were more closely spaced, with a higher probability of infant mortality.

In contrast, cultures that adopted livestock production gained the means to provide enough milk to ensure the survival of nearly every child, as long as they kept their rangeland forages from being depleted. Whether or not you are genetically predisposed to lactase deficiency depends upon how recently your ancestors adopted livestock and adapted to a novel set of nutritional opportunities associated with milk cows, goats, sheep, or water buffalo.

I once exchanged perspectives on this issue with food psychologist Paul Rozin, discussing his pioneering work on the significance of cultural selection for lactase tolerance. I found Rozin in New York City, where he was taking time off from teaching at the University of Pennsylvania to devote a full year to research at the Russell Sage Foundation. A man of modest build but commanding presence, Rozin had studied cultural culinary practices on several continents and had helped his former wife, Elisabeth Rozin, articulate a popular theory of "ethnic flavor principles" that underpin the world's major cuisines. But what Rozin and I spoke about that day was the peculiar manner in which *cultural* selection of ethnic diets has at times overridden innate biological tolerances to trigger genetic adaptation to new foods. In most cases, we think of biology *dictating* the path that cultural food preferences follow; that is, the natural selection of certain genetic traits tends to override cultural behaviors that do not always have immediate survival value. But, as Rozin has convincingly argued, "The biology-to-culture arrow can be reversed. Although we do not know the [historic] details of the pathway, the end product—lactose tolerance under genetic control—suggests that cultural practices of drinking raw milk and dairying provided the selection pressure for genetic change. Therefore, it is possible to go from culture to biology" (P. Rozin 1982).

The revolutionary significance of Rozin's evolutionary interpretation seems paradoxical at first, but its ultimate significance has not been lost on others. In his best-selling exploration, *Genome*, science writer Matt Ridley explained it this way: "The evidence suggests that such people took up a pastoral way of life first, and developed milk-digesting ability later in response to it. . . . This is a significant discovery. It provides an example of a cultural change leading to an evolutionary, biological change. The genes can be induced to change by [several generations of] voluntary, free-willed conscious action. By taking up the sensible lifestyle of dairy herdsmen, *human beings created their own evolutionary pressures*" (Ridley 2000; emphasis added).

Those ethnic populations that created their own evolutionary pressures, in this case, had to possess in low frequencies the gene for lactase production to begin with. But as long as this gene could be found among them, a relatively rapid rise in the frequency of this gene would occur as long as cheese or yogurt eaters gained nutritional and reproductive benefits from adding milk products to their diets.

The ultimate reasons that Gabriel and the Ak-Chiñ baseball team were inclined to used powdered milk for marking infields rather than drinking it are buried deep in the genetic and cultural history of their people. Until very recently, their bodies were shaped by hunting and gathering in an unpredictable desert environment rather than by herding cattle or sheep on the open range. Lactose intolerance is one of the ghosts of evolution encoded (then hidden) in their bodies.

~ Another ghost, a scarier one, is also present among the Pima, and Gabriel was also the person who taught me about the dark side of this dance between genes and drink. Until I was shaken by Gabriel's untimely death, I had not thought very much about how food and drink differentially influenced individuals of ancestry other than my own. Because Gabriel was the first Native American I had ever worked with side by side, day in and day out, I have deeply grieved his loss from this world. Ever since his death, it has been hard for me to drink or eat the things we once shared without his image appearing before me: long, straight, thick, raven-black hair; a mischievous, rounded face; thick forearms; and a barrel chest. Even if I had not known his ancestry, I still would have loved his riotous sense of humor, his throaty laughter, his unflagging allegiance to family and friends, and the heartfelt ways he shared his homeland with newcomers.

Gabriel was from a family of Pima Indians and grew up in the same Gila River Indian community as the Iwo Jima hero, Ira Hayes, who died drunk in an irrigation ditch a few years after World War II. Like Ira Hayes, the interaction between Gabriel's genes and his cultural and physical environment made him unusually susceptible to alcohol and to adult-onset diabetes, the latter affliction being prevalent among half the adults living in the Gila River Indian community today.

And yet, I had hardly noticed these vulnerabilities while Gabriel and I were busy building fences, shoveling manure, and planting vegetable crops for elderly Native Americans living on a desert Indian reservation near the one on which he grew up. Remembering those days, I wonder how much he or I were even aware of differences among people back then. After all, both of us were in our twenties, at the peak of our capacity for physical endurance, and so we behaved as if we were equally invincible. We worked hard renovating fields and gardens all day long and played hard in the evenings, going out to all-night "chicken-scratch dances" where Indian bands played endless polkas, cumbias, waltzes, and boleros while we swirled around the dance floor with our partners. Before dawn, we would devour bowls of *chile colorado con carne*, piles of pinto beans, huge flour tortillas or fry bread, and then wash it all down with a beer or two.

We justified our enormous appetites by talking about all the back-breaking labor we had been doing, for as it approached the harvest, we would have been working double time. Although both of us were already somewhat overweight, we stayed so physically active that we assumed we were leading healthy lives. Because we occasionally helped the reservation's nutritionists with village workshops on growing native foods to prevent diabetes, I knew that Gabriel was familiar with the nutritious foods that formed the basis for his people's traditional diet. Even though they were being abandoned by many of his kin, I knew that he still had access to them. During those first couple years of working with one another, I certainly did not worry that Gabriel would become a diabetic or an alcoholic.

What became belatedly apparent to me was that Gabriel's good personal intentions and family-oriented instincts were ones that could be easily derailed. A few of our mutual friends on the rez were prone to binge drinking, and sometimes Gabriel would join them, disappearing for several days. I would try to listen quietly, nonjudgmentally, whenever he returned to work hungover and disheartened, having to deal with the problems that the binge had created for him at home and in the office. Once, when he had avoided such perils for several months running, I invited him to a celebration at my home. He came early to help me set up tables, chairs, and coolers. When we about had it all ready to go, he pulled me aside so he could say something before the others began to arrive.

"Hey, buddy, . . . I . . . I hope you don't mind watching out for me tonight."

"Watch you? Watch you what? Watch you dance your way into cumbia heaven?"

"No, man, I mean watch me if I start drinking too much or wolfing down the food. . . . I get hooked before I know it, you know, so watch out. . . . Hell, I don't even know how to explain it to you. Well, what I mean is, even though you hang out with all of us Indians, I don't think you know how *different* it is for us." He grew sad, his voice tapering off, "Sure, you stay up with us all night, but I just don't think it hits you the same way."

I was suddenly aware of some palpable distance that had edged between him and me, a distance I had also felt between friends I had back in town and the rez gang to whom Gabriel had introduced me. "You mean the way *alcohol* hits you?"

"It's not like getting a little buzz off a beer, you know. . . . For us, it's like going into an entirely different space . . . we're in there by ourselves, not ever wanting to come out. Hey, I'm not trying scare you—as if I'm gonna to get drunk and rowdy tonight—but just keep an eye on me, OK? I mean, it's hard for me to even talk about. . . . Look, you need anything else done? You know, before all our buddies show up?"

Gabriel did fine that night without me looking after him much, but within months, he was dividing his time between parties, hangovers, and the hospital. Over that final, nightmarish year of his life, he was frequently treated for liver problems, extraordinarily high blood-sugar levels that worsened his diabetes, and, if I remember correctly, an ulcerated stomach lining—sometimes triggered by excessive drinking. The last time I saw him he was hitched up to a bunch of tubes and electronic monitors in a small, shabby, understaffed clinic run by the Indian Health Service. During his last period out of the clinic, he had cut his long hair down to the nubbin, and his weight had dropped precipitously. There was now an air of resignation about him—not only was he continuing to have health problems, but one of his teenage daughters had run into difficulty as well.

"*Shap a'i masma, ñ-nawoj*?" I asked him in his native language, not knowing that "How you been, buddy?" was the last question that I would ever be able to ask him about anything.

"I don't know. I don't know how much longer I'm gonna be around, either . . . ," he paused, his mouth dry. "Do you remember what I said to you one time? You know, that time when there was that party at your house? Remember? That it's different for us desert Indians. Whether it's alcohol or diabetes or both, it just hits us harder, 'enit?"

I could not stand to listen to him talk that way any longer. "Hey, buddy, *lighten up*," I said, now crestfallen myself. I struggled to cheer him up, but it all came out too objective, too analytical, too preachy: "This ain't your fate. And anyway, there's alcoholism in my family too . . . in a lot of families of all kinds of people. Being hospitalized don't mean that your body's gonna go down the tubes for good. . . . "

"Just listen to me, okay?" he replied. While his words softly flowed out of his mouth, they carried a pain that was deeper than any I had ever wanted to feel. "I'm trying to tell you something: this stuff is killing me . . . it's killing my people. I'm just asking you, what I've asked before: Why's it so different for us? Why do so many of us go down like this?"

He turned away from me, rolling his head toward the wall, and I never looked into his eyes again. My own were streaked with tears. I left the room, ducking out the clinic's back door. I walked around in the desert behind the clinic until my eyes were dry once more. It was not too long after that that Gabriel was gone for good.

~ After Gabriel died, I tried to stay in touch with his wife for a while. I wanted to do something to remember him, so each year on the anniversary of his death I presented the tribal library with books on native foods, gardening, and health in his name. It still left me feeling hollow though, like I had somehow failed him. At the same time, his death sapped me of any interest in drinking for some time; I did not drink even a drop of alcohol for more than a year. At another, even more irrational level, I was mad at Gabriel, feeling that he had "given up the fight."

What fight? The fight against stereotypes about "drunk Indians," "fat lazy Indians who can't keep their jobs," and "Indians who eat so much junk they've all become diabetic." I hated the idea that those who did not know him well would reduce his life to one of these ugly stereotypes. Of course, I knew plenty of Native Americans who did not in any way abuse alcohol, who were fit, and who blended into their diets many of the same nutritious, savory foods that their ancestors had eaten.

Perhaps underlying all my other responses, I wanted to know more about the notion that Gabriel had exposed me to: that just like their trouble with milk, many members of his ethnic community seemed to suffer more devastating consequences from consuming fast foods and alcohol than did other Americans. Empty calories are never benign, but for some more than others, they are immediately malignant. For certain individuals or special populations, even low or moderate concentrations of alcohol or simple sugars may cause physiological problems of a magnitude greater than what others suffer.

While the medical profession often labels these vulnerable populations as "genetically predisposed," in reality, their susceptibility to diabetes, lactose intolerance, or alcohol is not so hardwired that it should breed a certain fatalism, as it did in Gabriel. Instead, there is an interaction between genetic, ecological, and cultural factors that makes these people susceptible to high concentrations of sugars, particularly fermented ones; it is also a susceptibility that they can sidestep.

Nonetheless, there is now ample evidence to confirm that Gabriel was onto something: various people do differ greatly in their physiological, metabolic, and psychological responses to concentrated sucrose and its fermented derivative, ethanol. In the case of the Pima Indian families that Gabriel came from, there is now good data showing that they are predisposed to alcohol dependence by specific genes on chromosomes 4 and 11, genes that control dopamine and alcohol dehydrogenase metabolism.

Speaking in more general terms, I can say with some confidence that the same consumption level of alcohol can produce pronounced differences among individuals and ethnic populations. It can differentially change their rates of alcohol absorption and ethanol degradation, their heart rates, the intensity of euphoria they experience when inebriated, and it can trigger varying degrees of dizziness, facial flushing, muscular debility, and abdominal discomfort.

While some of these differences are correlated with gender, body mass, time of initial exposure to drinking behaviors, and to interactions with additional ingested substances, other differences are strongly heritable. Sensitivity or tolerance to alcohol is influenced by at least eight genes, many of which are polymorphic. In other words, there are genetic variants known as alleles that can either enable or disable the enzyme production of alcohol dehydrogenase and aldehyde dehydrogenase in the liver, the organ where alcohol is oxidized. Those individuals who produce a lot of these enzymes tolerate modest to heavy doses of alcohol, whereas those who do not can be either highly vulnerable to drunkenness or so sensitive to alcohol that they quickly learn to avoid it.

We now know that many (*but not all*!) Native Americans and Asians respond more severely to drinking a mild dose of alcohol than European Americans given the same dose. The percentage of people in Native American, Taiwanese, Chinese, Korean, and Japanese populations suffering strong reactions to low doses of alcohol can be as much as five to eight times as high as the percentages among European and European American populations.

Curiously, ethnic populations that *overproduce* alcohol dehydrogenase enzymes in the liver—thereby conferring a greater tolerance to fermented beverages—also have long histories of residence in regions where irrigated agriculture and livestock production was anciently practiced. These regions are also where there has been long exposure to dysentery resulting from unclean drinking water, contaminated by microbes associated with livestock and human feces.

As Matt Ridley has hypothesized in his groundbreaking book, *Genome*, these agrarian populations may have reduced their exposure to dysentery by drinking fermented beverages made from grains, grapes, or potatoes, instead of drinking untreated water. Nomadic people—such as Gabriel's ancestors who, until four centuries ago, obtained more than half of their calories from wild desert foods—had, until recently, little exposure to livestock-fouled drinking water and hence little hygienic incentive to produce fermented beverages year-round. At most, they fermented the juices of cactus fruit and century plants into rich, nutritious beverages for the briefest periods every summer, but these people went entirely without distilled beverages. Though common now, such potent drinks are still as difficult for their bodies to absorb as cow's milk is, for such beverages are truly foreign, nearly toxic substances when placed in the context of these people's evolutionary history.

~ In many ways—both brutal and subtle—Gabriel's vulnerability to simple sugars, milk, and alcohol were shaped by the interactions between his genetic heritage and the desert landscapes where his forefathers hunted, gathered, and searched for scarce water. I have already offered you a similar statement to describe the metabolic preferences that each of us has for certain cuisines: *we are what our ancestors drank and ate*. The longer the chain of ancestors who lived in one place—exposed to the same set of food choices, diseases, and environmental stresses for centuries—the greater the probability that selection was both for a diet and for genes that worked well in that landscape. The less that our ancestors intermarried with individuals from other lands, the greater the probability that we still carry genes that allow us to survive, thrive, and successfully reproduce under those particular environmental conditions.

Call this deep-time pressure on our diets *evolutionary gastronomy*. Paul Sherman, a behavioral ecologist at Cornell University, earlier termed it *Darwinian gastronomy* (see chapter 5), but Darwin's bias was that genetic changes in populations always took place over considerably lengthy periods of time. In that sense, I am not a Darwinian, for it has become clear that *microevolution* can rapidly lead to significant genetic change and divergence among populations in a matter of a few generations. From Darwin's finches to toads in the Caribbean, measurable change in animals' morphology, anatomy, and behavior are now known to have occurred within the sight (and lifetimes) of a single cohort of biologists. Under intensive selection pressure, faunal populations have differentiated into distinct subspecies and populations in a matter of a few generations.

Others have called this emerging field *nutritional anthropology*, while a few scientists see it as a subset of chemical ecology—the study of how secondary compounds affect food chains. If Darwin could hear such jargon-laden terms for basic life processes, he would roll over in his grave! And yet, Darwin would immediately recognize that imbedded in such terms is a wondrous hypothesis: that there is something profoundly functional in the mix of ingredients, cooking techniques, and preservation strategies characteristic of each ethnic cuisine, *for each traditional cuisine has evolved to fit the inhabitants of a particular landscape or seascape over the last several millennia*. Nutritional anthropologist Solomon Katz believes that this field offers us altogether fresh insights about our bodies and our tastes, seen as reflections of the evolutionary interactions between cultural diversity and biological diversity:

> The modern study of the origins and range of variation of human diet is directly affecting our understanding of human evolution. What humans eat is largely dictated by cultural traditions, but the degree to which a diet satisfies basic nutritional needs largely depends on . . . biology. This

> obvious interface between biology and culture has encouraged the development of a new approach or "paradigm" that analyzes and interprets biological and cultural adaptability as continuously interacting phenomena through human evolution. . . . [It is now clear that] human populations have biologically evolved adaptations to specific plants and other foods on which they have become dependent over very long time periods (Katz 1990).

What is also clear—perhaps for the first time in human history—is that the forcing of ethnic populations to abandon either their homelands or their traditional diets has inevitably led to epidemic rises in diabetes, heart disease, cancer, and allergies, among other maladies. Because some people have been untethered from the foods to which their metabolisms are best adapted, *some 3 to 4 billion of your neighbors on this planet now suffer nutritional-related diseases*. Those diseases—and how we can prevent them—are essential threads that hold this book together.

And yet, I must foreshadow this story by adding a sobering note. Do not assume that the future will bring us effective controls for diseases by producing gene therapies made possible through the new eugenics movement. Instead, the cautionary tales that follow will direct you back to what is in your ancestral garden and on your home plate, not to what lingers undiscovered in the test tube. We may use genetic research to better understand how particular genes interact with specific foods and other features in our environment, but the solutions will not be quick genetic fixes. Instead, more precise dietary guidelines and more holistic prescriptions for healthy living will likely be what is in store.

It is even doubtful that the latest rash of "nutriceutical foods" will be what allows us to live longer, healthier lives, although, by some accounts, these nutritionally-designed food products now make up about half of what is offered in American supermarkets and health stores. As you delve into evolutionary gastronomy in greater depth, I am sure you will see why such "new cuisines" cannot be so easily thrown together and mass-marketed in ways that will ensure that every one of us will achieve optimal health. The extensive pleasures as well as the psychological and physical staying powers associated with traditional cuisines are not coincidental; they have been carefully worked out over history—evolutionary history, *our rich and varied history*. They have been shaped by the diversity of cultures that have hunted, herded, farmed, and foraged on this Earth. Moreover, these traditional cuisines have helped to foster human diversity in astonishing ways that have only recently been elucidated by scientists such as Fatimah Linda Collier Jackson, an African American who studies human genetic interactions with traditional African crops (such as sorghum and cassava) from biological, anthropological, and nutritional perspectives. As Jackson has remarked, "Contemporary human biological diversity may reflect the differential exposure of various ancestral and modern groups to diverse environmental constraints, including variable exposure to plant-derived secondary compounds . . . it is unlikely that human dietary contact with plant [and animal] chemicals has been without biological and behavioral consequences. More likely, chronic exposure to specific plant compounds has been a salient part of our species' evolutionary experience" (Jackson 1991).

It is this *chronic*, persistent engagement with particular plants and animals prepared for eating in very specific ways that food historian Elisabeth Rozin claims is the essence of ethnic cuisines: "Culinary behavior, or what we more commonly call cooking, is practiced not just occasionally or under special limited conditions, but with a frequency and a regularity that are true of very few other activities. Yet, while all people do it, they all do it differently. . . . People who define themselves as a group express or interpret the general human practice in their own terms, and it is this style or expression of universal culinary activity that we call *cuisine*" (E. Rozin 1989).

This holds true for the earliest hominids scavenging on the African savannas, just as it does for Arctic hunters of blubber-laden marine mammals and for rain-forest dwellers who learned how to detoxify poisonous cassavas. The practitioners of each enduring cuisine have offered both sensory pleasures and tangible health benefits to those at their campfires or tables—pleasures and benefits that are peculiar to a particular place of origin.

Remarkably, many of these olfactory and nutritional rewards are fine-tuned to various cultures' genetic legacies in mysterious and miraculous ways. For me, at least, they seem miraculous because they tell the fundamental stories of human origin and divergence in ways that few of us have heard before. It is this corpus of stories taken from my time among many cultures scattered around the world that form the underpinnings of this book. As Elisabeth Rozin again reminds us, such stories are actually all around us, every day; we simply need to learn how to listen for their sheer melodic beauty and harmonic resonance:

> In the dusty streets of rural villages, in the dingy rooms of city tenements, in the furtive clearings of sweating jungles, in the secret, sacred precincts of three-star restaurants, it is going on. Listen, and you will hear it: the clatter of pans, the slapping of dough, the pounding of grain. Sniff, and you will smell it: the roasting meat, the newly baked bread, the aromatic sauce. Look, and you will see it: the quick stirring of a pot, the delicate folding of a triangle of dough. . . . Wherever you may wander, among those of humankind, you will find them preparing their food, expending enormous quantities of time, energy and attention, on that homely activity we call *cooking* (E. Rozin 1989).

CHAPTER TWO

# *Searching for the Ancestral Diet*

## Did Mitochondrial Eve and Java Man Feast on the Same Foods?

THESE DAYS, it seems as though eating is done with more self-consciousness than ever before in human history. Most everyone is trying out some "miracle diet," one whose champions claim will keep them fit, prevent diseases, make them look sexy, and ensure greater longevity. Nearly everywhere we look, there are TV doctors and nutritional-supplement hawkers, cookbook divas and sports celebrities trying to hook us on a diet plan that they claim will cure all that ails us. Many of these miracle diets are based on trendy theories about what is nutritionally best for *the* human body in this day and age, given our exposure to an unprecedented set of toxic additives and highly derived compounds such as trans-fatty acids. While a few of these dietary theories divide all of humankind up into a few groups with different nutritional needs based on a handful of blood types or metabolic inclinations, these diets are the exception rather than the rule. Indeed, the vast majority of these miracle cures tend to gloss over the ways in which our bodies and our genes are different from one another. At present, my body may look like the fun-house mirror distortion of your body, but once we both start eating the latest version of the optimal diet, its proponents claim that we will look the same.

Curiously, some proposals for what the optimal diet may be sidestep the peculiar challenges of staying healthy in the current tech-nonindustrial world while encouraging us to delve into the past; these diets want us to remember our bodies' inherent capacities and our minds' predilections for foods that have been shaped over evolutionary time. As nutritional anthropologist Boyd Eaton and his colleagues have recommended for some two decades, we might choose what to eat by paying more attention to the "Paleolithic legacy" written into our own blood and bones, nucleotides and genes. By Eaton's reckoning, the very foods that our hominid ancestors ate millions of years ago are still what our metabolisms are best suited to consume.

This evolutionary perspective was later incorporated into a landmark paper, "The Dawn of Darwinian Medicine," in which ecologist George Williams and psychiatrist Randolph Nesse blamed most present-day nutrition-related diseases on the fact that we no longer eat and exercise as we did during the period of human origins. We ignore at our peril, they claim, the fact that "human biology is designed for Stone Age conditions."

As one of Eaton's disciples, paleonutritionist Loren Cordain has written in his popular cookbook, *The Paleo Diet*, that an ancestral cuisine is "*the one and only diet* that ideally fits our genetic make-up. Just 500 generations ago—and for 2.5 million years before that—*every human on Earth ate this way*. It is *the* diet to which all of us are *ideally suited*, and *the lifetime nutrition plan* that will normalize your weight and improve your health. I didn't design this diet—Nature did. *This diet has been built into your genes*" (emphasis added).

There are now millions of people around the planet who agree with Eaton, Williams, Nesse, and Cordain, and they are attempting to locate and eat the very foods they presume to be most fitting for their Stone Age–designed bodies. Every day, hundreds of thousands of people follow menu plans found in a set of books that collectively inform the "ancestral diet" movement. Most of these menus are based on the tenet that there is but one diet that fits the human metabolism and that is the diet most akin to what our ancestors ate during the earliest eras of human evolution.

Going as far back as we can toward our common human origins gets many of us off the hook from identifying a single ethnic diet derived from our more recent ancestors. With so much ethnic intermarriage these days, many of us grew up exposed to the comfort foods of several different cultures rather than sticking with one cuisine that jived with some simplistic notion of pure-blooded ancestry. Statistically speaking, most of us are mutts rather than blue bloods, so that it is getting ever harder to select one ethnic diet that may speak most directly to our genes, as *the* diet to which our metabolism is hardwired. This dilemma is especially evident for the 7 million Americans who identify themselves as composites of two or more "races"—whatever a "race" is considered to be today.

Accordingly, it may be more comforting for most of us to eat our way farther back in time, loading our plates with the very same foods that our great-great-great-etc.-grandmother Lucy once served in her camp near the Olduvai Gorge thousands of generations ago. Hundreds of thousands of dieters have chosen to do just this, pledging to spend their budgets on calories, cures, luncheons, and literature that pursue a Paleolithic prescription, one that ignores *ethnicity* in exchange for a sense of *antiquity*. They have become hooked on Website versions of an ancient cuisine variously referred to as the Cave Man Diet, the NeanderThin™ formula, the Origins Diet, the Stone Age Menu, the Paleo Diet, or the Carnivore Connection.

Dieters following these plans are romantically attempting to reconstruct just what exactly it was that Eve, Lucy, or Java Man may have eaten around the campfire in the olden days, with a few food-preparation shortcuts thrown in. Unfortunately, if you would prefer recipes that are authentically grounded in knowledge derived from paleonutritional, zooarchaeological, and ethnobotanical studies, there are few signposts along the trail to assure that you are truly eating your way back to your roots.

Just what are the contemporary foods acceptable as ingredients in "ancestral" diets that have taken on the aura of lifetime nutrition plans? Can a descriptive reconstruction of past diets be used in a prescriptive manner for today's health problems? Ask a dozen nutritionists, physical anthropologists, and paleoethnobotanists, and you may get two dozen answers. Nevertheless, most proponents of ancestral diets do agree with some basic parameters that I will try to distill from the writings of the more legitimate scholarly sources:

1. Our hunter-gatherer ancestors may have gained as much as 65 percent of their energy from vertebrate animals, eating all parts of game and fish, while seldom eating eggs and never consuming milk products.
2. In addition, these foragers consumed in raw forms a variety of fresh fruits, flowers, leaves, and bulbs, many of which are rich in disease-preventing compounds that have since been bred out of most of our cultivated food crops.
3. Our ancestors rarely ate any quantity of cereals and certainly did not finely grind grains and other small seeds into fiber-depleted flours.
4. Neither did they consume quantities of sodium salts, although their diet was rich in calcium and potassium salts.
5. As nomads, our ancestors seldom camped in any single place long enough to let fruits or other carbohydrate-rich plant parts ferment into ethanol or acetic acid (vinegar), and they certainly were not involved in the distillation of highly potent alcoholic beverages.

Of all the paleonutritionists, perhaps Eaton himself takes the longest view of dietary evolution, emphasizing that "the nutritional requirements of contemporary humans represent the *end-result* of dietary interactions between our ancestral species and their environments extending back to the origins of life on earth" (Eaton et al. 1996). Nonetheless, Eaton and his disciples have paid particular attention to how our hominid ancestors foraged in wild habitats around

2.5 million years ago. They claim that our ancestors' foraging patterns continued essentially unchanged for hundreds of thousands of years, shifting substantively only when farming and livestock raising began some 10,000 to 12,000 years ago.

But if we are to get down to brass tacks—by precisely defining in some detail what our common ancestors ate that may still be available and palatable to our modern sensibilities—we must gain some sense of whether there was ever much place-specific variation in dietary preferences. In other words, what our single common ancestor (nicknamed "mitochondrial Eve" by geneticists) ate at one spot in ancient Africa over a lifespan of twenty or thirty years really is not the issue; it is how much our other ancestors' diets deviated from her food choices and whether the dietary variation among them had much influence on the design of our bodies.

My hunch is that most characterizations of ancestral diets woefully simplify such variation, ignoring incredible levels of dietary diversity that have guided our evolution in space and in time. So instead of searching for *the* ancestral diet only where the oldest human remains occur in the Rift Valley of Africa, let us sail all the way over to Bali and Java, among the farthest places from Africa to which protohumans strayed long, long ago. These are places that presently fall within the nation of Indonesia, and ones that have been the haunts of a wild-food forager who goes by the name of Java Man.

~ Java Man was named by Eugene DuBois, an adventurous scientist who first made his way from the Netherlands to Java and Bali in 1880, when the islands were still parts of the Dutch West Indies. DuBois knew that orangutans still survived in the mountains of Java, and so he speculated that the uplands might also harbor skeletal remains of links between their history and ours. He followed several fruitless leads until 1891, when he excavated an ancient skull that suggested he was finally on the right track. It was not until the next year, when he found a thigh bone and a few teeth, that he proclaimed he had indeed found *the* missing link.

In retrospect, it does appear that the skull belonged to one now-notorious individual of *Homo erectus* (a.k.a. Java Man) who lived 800,000 to a million years ago. The femur and the teeth, unfortunately, have been determined to belong to two other species. At the time of its discovery, DuBois had proclaimed that he had uncovered the femur of a primate who had walked around Java in an upright position; that particular fossil is now regarded to be the remains of a human individual who lived much later than Java Man. And the teeth . . . well . . . they do not belong to either Java Man or Mr. Upright; instead, they appear to have served a now-extinct orangutan species in his masticatory pursuits. Nonetheless, the highlands of the East Indies had proven to be fertile ground for studies of our evolutionary history.

When I first traveled to the East Indies, I was constantly trying to determine what foods Java Man may have been exposed to that might still occur on the islands, and what might have changed in the meantime. Within hours of landing on Bali, I was able to travel by bus to a tropical beach where I could scan the eastern horizon for the coastal cliffs and volcanic summits of Java. I recognized that dramatic changes to these islands had occurred over the millennia—not merely affecting their shapes, sizes, and seasons, but their floras and faunas as well. As my hosts offered me a sampling of some flavors characteristic of Bali, I felt as though I were tasting foods more akin to what DuBois' Java Man had eaten than to what mitochondrial Eve had ever tried in the Rift Valley of Africa—snake fruit, water apple, melinjo, and rambutan, as well as smoked reef fish wrapped in pandanus leaf.

As I experienced my first dawn within this tropical archipelago, someone pointed out to me the last of some flying foxes coming back to roost in the palm fronds above our heads. Larger than any bat I had ever seen in the Americas or in Africa, the presence of these unique animals made me realize that I—like Java Man millennia before me—had reached into a Southern Pacific realm where a distinctive biota had evolved. Java Man is among the best documented of our hominid ancestors who ranged well beyond the forests and savannas of Africa. He must have foraged across a wide range of habitats, including those once associated with the Java land mass when it was connected to Eurasia by a land bridge, and neighboring Bali when it was emerging from the ocean. Did he eat the same things everywhere? Did his diet change through time?

Sourcing a list of ingredients for a reconstructed ancestral diet to match that of Java Man will, of course, prove difficult. When Java Man's species, *Homo erectus*, took up residency on the land mass that survives today as Java, giant land tortoises, pangolins, and mastodons were among the many large mammals and reptiles that roamed the coastal lowlands. This megafauna was no doubt part of human diets, at least for a while, but on most islands where such giants formerly roamed, humans sooner or later had a hand in their demise. Today a third of the species known to be contemporaneous with Java Man are globally extinct; another third occur elsewhere, but not on Java nor Bali; and the final third (including flying foxes) persist, but in relatively low numbers on these islands.

It is this latter third that interests me, for it indicates that Java Man may have been familiar with some of the very same animals and plants present in today's East Indies. Surprisingly, some of the species are edible ones that I, like pioneering biogeographer Alfred Wallace a century before me, had previously sampled on other land masses. The plants may not grow today in the exact places that Java Man or even Wallace found them, but perhaps they are still within grasp.

Out of the bus window on Bali, I began to spot familiar trees that I had also seen grow wild in parts of Asia and even Europe. The presence of these ubiquitous plants did not mean that they had recently invaded Bali; to the contrary, in prehistoric times, most had naturally been dispersed to Bali, Java, and other islands close to the Asian mainland. This floral exchange with Asia, Europe, and even Africa was what fascinated Wallace, one of the earliest and greatest scientific explorers of the Balinese highlands. Yes, this was the same Wallace who fleshed out the theory of natural selection almost as if it were a malaria-induced hallucination, which came to him in his sick bed, while the more methodical Darwin labored away on the same theory for some thirty years.

Wallace was an astute observer of natural variation and similarity; he was quick to recognize that the berries and wildflowers he spotted in the mountains of Bali were of the same species that he had first learned in the hill country of England, or later, on his holidays to the Alps. Curiously, the edible species that Wallace found in the highlands and on other islands of Indonesia have implications for our understanding of the dietary breadth of early humans. Given the degree to which the northern islands of Indonesia have shared elements of their flora and fauna with Africa and Eurasia, it is not surprising that *Homo erectus*, and later, *Homo sapiens*, spread this far east, encouraged by some initial dietary choices that could have superficially appeared to be familiar to them. After all, any plum looks and tastes somewhat like any other plum, regardless of whether it grows on the banks of the Upper Nile or on the beaches of Bali.

But as I soon learned by venturing out by boat to islands south of Bali, the continuous chain of plant foods that I could find from Africa and Eurasia to Bali went no farther. While my botanical training helped me quickly pick out the few plants on Bali that were shared with other places I had traveled to, it also prepared me to recognize the huge turnover in plant species that I would see between Bali and the islands reaching south to Australia.

It was Wallace who first grasped the full significance of this discontinuity, although he was not thinking of its significance to human diets. I decided to loosely retrace the trajectory that Wallace had taken more than a century ago, traveling to some islands with plants that look (and taste) very different from those on Bali. With friends, I chartered a sailboat and headed southward toward the islets of Lombagan and Penida, only ten miles off Bali's shores. If paleoanthropologist Mike Morewood is correct, Java Man and his kin could have now and then made such a journey, establishing satellite populations of *Homo erectus* on neighboring islands: "*Homo erectus* was not just a glorified chimp. . . . We now believe they made sea crossings to reach Flores and other Indonesian islands" (Morewood 1997).

What I could see on Lombagan, just as Wallace had seen on several islands south of Bali and Java, was how abruptly floras as well as faunas could change in a matter of just ten miles, and accordingly, how dramatically the human diet must have had to shift as it adapted to newly occupied habitats. Inevitably, there must have been significant turnovers in the composition of ancestral diets as our progenitors moved through space as well as through time. It was simply not possible for the historic dwellers of Bali and Lombagan to share much of the same diet, optimal or not. In fact, my visit to Lombagan and other outliers convinced me that very few foodstuffs could have been shared by ancestral peoples who contemporaneously ranged from Africa to what is now the East Indies, since few foods were even shared between folks on the coasts of Bali and Lombagan—people who lived near enough each other to send smoke signals back and forth.

Here's why. Lombagan and the neighboring isle of Penida are decidedly arid, not tropically moist like most of Bali. Thorny, little-leafed acacias, sennas, mimosas, and straggly tree euphorbias line their coasts, while their inlands are far too small and dry to raise rice-like grasses at all. They are covered not by the lofty, massive canopies of dark green leaves found on Bali, but by dozens of grayish, ground-hugging aromatic shrubs—ones loaded with terpines and other aromatic oils. Most of these plants have small, leathery, water-conserving leaves and oily berries that are miniscule in comparison to the hundreds of large tropical fruits I saw in the Balinese highlands.

When I went ashore on Lombagan, I felt caught between the glaring sun and the reflected heat rising from the rocks at my feet. Compared to the way I had nestled deep into Bali's cool, multilayered shade, on Lombagan I felt more exposed, more thirsty, and even prone to heat stroke. It made me visually hungry for Bali's almost excessive abundance of everything green and edible.

It also made me remember Wallace's Line, a tangible example of real geographic barriers that have made the food resources available in one place vastly different from those in another. Wallace had discovered this biogeographic border while venturing from Bali to Lombok, another island to the south that is even more eerily arid than Lombagan and Penida. Wallace's Line is an ancient barrier that effectively halts the further dispersal of plants and animals between the Eurasian and the Australian biogeographic provinces. Whereas Bali and Java share some 97 percent of their bird species, the island of Lombok—some two dozen miles south—shares only 50 percent of its birds with Bali. When Wallace compared Lombok to Bali's lush, leafy, fruit-laden forests, he described its depauperate thornscrub as "a parched-up forest of prickles . . . the bushes were thorny, the creepers were thorny, the bamboos were even thorny . . . everything grew zigzag and jagged in an inextricable tangle" (Wallace in Van Oosterzee 1997).

*An inextricable tangle*: that is exactly what happens among the genes, foods, and habitats of plant, wildlife, or human populations when they become even somewhat isolated from other populations in different habitats nearby, whether they live on different islands or at various elevations on the same land mass. Wallace demonstrated that there are hugely different selection pressures placed on plants and animals inhabiting distinctive landscapes just a few miles apart. Because such populations exist in some degree of reproductive isolation from one another, a second evolutionary mechanism favoring divergence, *genetic drift*, is also active. Genetic drift is the skewing of the frequencies of genes in populations that becomes more pronounced the smaller and more isolated those populations happen to be. Thus, once a gene for dwarfism is introduced to or emerges within an island species, it has a higher probability of spreading and becoming dominant in the small population than it would in a larger population.

And so, most islands have what we might call a skewed set of foods containing a skewed mix of dietary chemicals in them. They are considered to be skewed when compared to what colonizers might have been familiar with in mainland habitats, thanks to the natural selection that has occurred in those dissimilar environments and the larger role that genetic drift plays in small populations. Darwin and Wallace were the first biologists to describe the consequences of plant and animal species' divergence that biologists now refer to as the process of *adaptive radiation*. As an insightfult observer of patterns in the natural world, Wallace in particular recognized the dramatic differences in the floral and faunal composition of islands hardly set apart from one another spatially. Compared to Wallace and other contemporary biogeographers, Darwin's vision was limited in a way that until recently was prevalent among students of evolution. While Darwin surmised evolutionary processes by making comparisons across the relatively short distances he traveled in the Galapagos, he still believed such processes could be observed only as their cumulative effects spanned great stretches of time or space: "It may be metaphorically said that natural selection is daily and hourly scrutinizing, throughout the world, natural variations; rejecting those that are bad, preserving and adding up all that are good; silently and insensibly working, *whenever and wherever opportunity offers* . . . [but] we see nothing of these slow changes in progress, until the hand of time has marked the lapse of ages" (Darwin 1859, emphasis added).

But what is now apparent—both from ecological studies of island plants and animals and from genetic studies of island peoples—is that many of the changes in genetic frequencies do not proceed as slowly as Darwin or even Wallace had assumed. As Harvard-trained science journalist Jonathan Weiner has so vividly elucidated in *The Beak of the Finch*, evolution is observable in "our time" and in "our species." Our bodies' responses to particular diets were not fully shaped 2.5 million years ago during the emergence of the genus *Homo*, nor were they fixed during the period when mitochondrial Eve roamed the savannas of East Africa between 150,000 and 200,000 years ago. They have been constantly reshaped by the peculiar range of food choices, environmental stresses, and diseases that humans face in every place in which they have spent considerable time; and, of course, our reactions continue to be reshaped by our present food choices and disease exposures as well.

~ Human populations like our own have encountered distinctive sets of foodstuffs and dietary chemicals that interact with our genes wherever we have lived. Perhaps that is the essence of why archaeologists and paleoecologists search through debris for bones of fish, fowl, and game, and for seeds, pits, and plant stems at places on Java or Bali where ancient human bones have been found. It is also why I spent time scanning the leftovers around far more recent human encampments on the beaches of Bali, Lomagan, and Penida. The more campfire rings I saw, the more I noticed what a motley mess of kitchen scraps humans tend to leave behind as trash wherever we camp. At one site, I saw plenty of fire-cracked clams and charred fish bones, but little else. At another camp a ways inland, there were rotting leaves and shoots of wild vegetables and a few tattered fragments of insect carapaces.

Each time that I have visited with archaeologists who sift through the soil at early hominid camps, they seem less certain that there is a single discernible dietary pattern evident among excavated sites. Some scholars have begun to doubt whether Java Man or other populations of *Homo* ever kept to a uniform diet; some even wonder if ancestral diets contained more or less the same proportions of fats, proteins, sugars, and fiber. Such a varied diet may simply have resulted because Java Man and his descendants craved different foods as they aged—

or became pregnant, sick, or injured—or as they increased or decreased their physical activity in work or in play.

And yet, there may be a deeper reason why Java Man may not have maintained any one dietary pattern, and accordingly, why there may not be an optimal diet for all humans, past and present. One explanation may be that we differ genetically from one another in small but significant ways that not only shape our food preferences, but that are reciprocally shaped by them. Human diversity—the genetic variation within our species—interacts with the diversity of edible plants and animals distributed across this planet in ways we are just beginning to understand. And while *diversity* of any kind is now celebrated in some social circles, its implications for how we eat have largely been ignored. Perhaps this is because of the many times in recent history that supposed physical or psychological differences among human populations have been used politically as a means to deprive one ethnic or "racial" group of opportunities that another "elite" group controls.

Now, however, with the unbelievably rich information suddenly made available to the public by the Human Genome Project, old scientific notions of races have lost much of their credibility. Of course, that does not mean that deep-seated cultural prejudices have suddenly vanished; like the ghosts of evolution mentioned in the previous chapter, they still haunt us whether we choose to see them or not. Nonetheless, it has become abundantly clear that differences in skin color are not well correlated with the way other human traits are distributed within our species. Moreover, eminent geneticists such as

Harvard's Richard Lewontin have conclusively demonstrated that more human genetic variation is situated within "racial" or ethnic populations than between these populations. By various calculations, differences between ethnic populations account for only 7 to 15 percent of our species' total heterogeneity.

And yet, much of the popular science literature on human genetics that I read continues to be full of simplistic truisms. One of those is that 99.9 *percent of the human genome is shared by each and every one of us and all of our ancestors*, regardless of how we self-identify our racial heritage. If I took this new cliché at face value, I would have to accept that all significant genetic characteristics of our species became "fixed" quite a long time ago, with few adaptive variations on any evolutionary theme persisting in the human genome. In other words, I would have to assume that the extant genetic variation found in humans living today is about the same as that which could be found among the range of human populations living one hundred generations ago—or for that matter, is the same as what we may find in other primates—as if there have been no new selective pressures on us for millions of years. If that were true, our contemporary dietary needs for a certain mix of macronutrients should hardly differ from those of our hominid ancestors.

The trouble with this argument is that vertebrate zoologists have found considerable variation in diet and genes even among primate populations of the same species living less than one hundred miles away from one another. There is also considerable genetic variation extant among all of us contemporary humans, and we survive and thrive upon a bewildering diversity of foods. So just how similar are our genetically determined needs for nutritional resources? Are they similar enough that we should all pattern our current diets after those of our common ancestors?

In the pages of the *New York Times Magazine* in 2002, Dr. Sally Satel—a physician with a penchant for delving into the philosophical dimensions of biomedical practice—tried to get a grip on the degree to which genetic determinism should guide us in accepting one-size-fits-all medical and nutritional recommendations:

> What does it really mean to say that 99.9 percent of our content is the same? In practical terms it means that the DNA in any two people will differ in one out of every 1,000 nucleotides, the building blocks of individual genes. With more than three billion nucleotides in the human genome, about three million nucleotides will differ among individuals. This is hardly a small change; after all, mutation of a single one can cause the gene within which it is embedded to produce an altered protein or enzyme. It may seem counterintuitive, but the .01 percent of human genetic variation is a medically meaningful fact (Satel 2002).

What does that mean with respect to our diets? It is doubtful that any set of recipes could be custom-made to meet *all of our* physiological requirements, given how much diversity there is among us. At best, a uniform dietary regimen could try to strike a balance of macronutrients catering to the 2,997,000 nucleotides that may have not changed much since our human origins, an approach that still glosses over our varying needs for micronutrients. At the same time, the mongrelized recipes in the ancestral-diet literature largely ignore the 3 million nucleotides that mark the divergence of all humans currently living from our common female ancestor, mitochondrial Eve, who lived 150,000 to 200,000 years ago. Since her lifetime, many millions of nucleotides have contributed to the diversity found among all humans past and present, 3 million of which continue to be expressed in the physical differences among those living today. Regardless of the intricate branching and intermixing of the human family into the great genetic diversity we see within our species, the recipes found in the ancestral-diet movement focus on the presumed *taproot* of our species, not on the branches. And it is the branches, not the roots, that may indicate how much nutritional needs of individuals and populations may vary instead of fitting a single pattern.

~ That single pattern—as the proponents of an optimal Stone Age diet claim in their cookbooks—is that hunter-gatherers nearly always consumed more animal foods than plant foods. Paleonutritionists insist that if you averaged it out over hundreds of thousands of years, the animal to plant ratio of energy intake would tip the balance with two-thirds meat to one-third veggies.

Ironically, when some of these proponents of Stone Age diets publish more-scholarly treatises in technical journals, they acknowledge how difficult it is to be precise about such ratios for most hunter-gatherer cultures, given the extreme variation that has been found among individuals, families, seasons, years, and habitats. When several of these scholars looked in detail at 229 hunter-gatherer societies, one in seven of these foraging cultures clearly consumed more plants than animal foods. In the *American Journal of Clinical Nutrition* in 2000, Loren Cordain and his colleagues concluded that "our data clearly indicate that there was no single diet that represented all hunter-gatherer societies."

Yes, you heard Cordain's name before, just a few pages ago. Remember what he stated in his popular 2002 cookbook: "The Paleo Diet is *the one and only diet* that ideally fits our genetic make-up. Just 500 generations ago—and for 2.5 million years before that—*every human on Earth ate this way*."

When I pointed out this discrepancy to him over the phone, Cordain defended this statement by arguing that there were indeed strong commonalities in the trends of macronutrient composition of diets documented among various hunter-gatherer societies. He did, however, agree with me that the micronutrient and secondary compound composition of hunter-gatherer diets varied greatly from landscape to landscape and season to season.

Micronutrients and other, secondary dietary chemicals are the spices of life that vary from place to place, time to time, and in ways that make the study of human diets so fascinating. Of course, wild plants and animals are astonishingly diverse in the chemicals that they contain, and these chemicals can both benefit *and* imperil our health. Many of the more toxic as well as the more protective chemicals that were formerly consumed routinely in wild-food diets around the world have been consciously and unconsciously eliminated by modern crop and livestock breeding, which is ever striving for more palatable, uniform products. But this winnowing away of the chemical diversity found in wild plants and animals had not yet happened when Java Man and our direct ancestors first spread across the face of the Earth; instead, our predecessors were fully exposed to the astonishing range of plant and animal chemicals found in various habitats. They learned what to forage (as well as what to avoid) from sea level to more than 10,000 feet in elevation on several continents. Each landscape that they entered harbored additional species, each with a different mix of protective chemicals and attractive nutrients.

As an ethnobotanist, I have helped document more than 350 species of plants historically used as food in the Sonoran Desert, a landscape not particularly known for high levels of diversity. On a global scale, however, some 30,000 wild plant species have been documented to have been prehistorically or historically consumed by various ethnic populations. Those plants are not merely sources of the proteins, sugars, and fats that Cordain and his colleagues recorded to use in their reconstruction of dietary ratios for an optimal diet. They are also arsenals of the potent chemicals that botanists call *secondary compounds*, plant chemicals that seem to serve no direct metabolic purpose in photosynthesis, growth, and reproduction. Instead, these chemicals—which comprise unique mixtures in different species and even in different populations of the same species—protect plants against environmental stresses such as drought, freezes, and fires, and from interspecific stresses such as competition, disease, predation, or herbivory.

When consumed by humans, these secondary compounds also give our plant foods their flavors and fragrances, their capacity to poison us or to kill us, their ability to protect against carcinogens or microbes, and their utility as aphrodisiacs or as fertility suppressants. Moreover, it has recently become clear that some chemicals in wild foods are potent enough to cause our own genes to mutate. Fatimah Linda Collier Jackson, the African American nutritional anthropologist mentioned earlier, has pondered the significance of this little-appreciated fact: that human genetic mutations can be induced by some wild herbs, legumes, and tubers that various human cultures have consumed since time immemorial.

As Jackson has documented, such plants' secondary compounds—especially those called *allelochemicals* for their capacity to reduce herbivory from visiting animals and competition from neighboring plants—can influence human genetic variation in myriad ways. Science writer Bruce Grierson has noted that "dietary chemicals change the expression of one's genes and even the genome itself" (Grierson 2003). He reminds us that not all dietary chemicals get metabolized as calories to fuel our work and play; some of them are transformed to ligands and attach to proteins, forming complex molecules that literally turn on and off the expression of certain genes. As a specific example, Grierson explains that a secondary chemical in soybeans known as genistein binds itself to estrogen receptors and regulates the expression of genes affecting hormonal fluxes. And yet, the consumption of genistein does not affect every woman's estrogen cycle in the same manner, since individuals from different ethnic populations carry different estrogen receptors that respond to genistein and other ligands to varying degrees. In short, there are complex feedback loops between preexisting genetic variability and the influence of secondary chemicals on gene expression and mutation.

With the accumulation of such evidence over the last decade or so, it has become increasingly clear that dietary chemicals are major driving forces for genetic expression, mutation, and selection within our species, not merely a sideshow. Jackson has boldly suggested that these secondary compounds have literally fostered human diversity through their inclusion in traditional diets over millennia.

Let us try to fathom how pervasive the influence of these chemicals in plant foods might be—and why numerous secondary compounds, not just macronutrients—have influenced the shapes of human diets and genetic variation through time. Some phytochemists have hazarded the guess that 40,000 to 50,000 secondary compounds have already been described from the 270,000 named plant species, and yet, not even a fraction of the 30,000 edible plant species have been subjects of substantive laboratory analyses. The chemists therefore concede that they have barely scratched the surface in characterizing the structure of these compounds, let alone in understanding their ecological function, their role in nutrition, or their mutagenic capacity (that is, their ability to induce genetic changes).

Nevertheless, there are some things we do know. Just one set of secondary compounds, the bitter-tasting alkaloids, can be found in one out of every five plant species, or some 54,000 of the currently identified species altogether. These species are not uniformly distributed around the planet; some landscapes, such as dry tropical forests, and some families, such as the nightshades (including tomatoes, potatoes, chile peppers, and eggplants), are more loaded with alkaloid-bearing species than others.

Many of the same secondary compounds can be mutagenic at one level of habitual ingestion, carcinogenic at another, and disease-preventative at still another. As many Latin Americans are well aware, a culinary herb like epazote (a relative of lamb's-quarters) can be a simple flavoring in a pot of beans, a reducer of flatulence if liberally added to the same beans, or an accidental abortifacient if a pregnant woman happens to ingest too much of the herb. Some wild populations of epazote are intentionally sought out and used by *curandera* herbalists to induce abortions in their human patients or in livestock.

Such properties are not restricted to a peculiar set of medicinal and culinary herbs. They are also found in fruits that are widely distributed around the world; particular plum populations may have their own potent mix of chemicals. Collectively, various wild plums contain as many as 150 known secondary compounds, but their concentrations vary among species and their populations. Some 67 of the secondary compounds identified in edible plums have been found to be bioactive, that is, capable of stimulating a variety of metabolic consequences that affect our health. All of these 67 compounds may be found in a particular kind of wild plum, but the different beach plums that cover the shores of Bali, Java, Hawaii, and Africa vary greatly in their concentrations of these compounds.

Throw the myriad kinds of plums on Earth into the same plum pudding, then observe the effects as we feed it to representatives of the thousands of cultures found around the planet, and we have what ecologists call biocomplexity. Now multiply that complexity by the 30,000 edible plant species on Earth, plus a few thousand edible animal species. It becomes exceedingly plausible that prehistoric Java Man was exposed to a vastly different set of compounds than were his contemporaries in Africa's Rift Valley. Java Man's species, *Homo erectus*, never covered even a fraction of the ground or achieved a fraction of the population size associated with our more recent species of *Homo sapiens*, and yet the evolutionary trajectory of *Homo erectus* was probably set by a dramatically different brew of secondary compounds than we consume today.

These secondary compounds likely fostered the considerable genetic variation in humans and in primates that I alluded to earlier. Although we may never be able to reconstruct just how much human genetic diversity has been lost through time, it is probable that there was far more genetic differentiation *between* the populations of our ancestors than there is today. If genetic studies of chimpanzee populations are any indication, there was also greater variability within each breeding population. Curiously, the chimps within one breeding population on a hillside in Africa express twice as much variability in their mitochondrial DNA than do all of the 6 billion humans currently living around the Earth.

By the time I had left Bali, I had become more aware of just how heterogeneous both our food choices and our own human populations once were compared to what we experience today. Despite the many plant and animal extinctions that have occurred within this last century, there remain tens of thousands of edible species within our reach, but most of us have narrowed our diets down to a few hundred domesticated species, and most of them have had their potent chemicals culled out by crop breeders. Since colonizing cultures began the spread of agriculture and the conversion of diverse wild landscapes, most human diets have become far more homogeneous than ever before in our species' history. We can only speculate how much more genetically narrow our species has become over the same time period of the last 12,000 years. When prehistoric human populations were widely scattered and exposed to so many distinct plant chemicals, perhaps the variation between such populations was on the order of two or three out of every six genes in the cumulative human genome. Without a doubt, there were greater genetic differences (as well as dietary differences) than we see among contemporary human populations, which only vary from one another in one out of every six genes.

As we shall see in the following chapters, livestock and crop domestication began to dramatically shape some human physiological responses as people adapted to new sets of foods. However, the emergence of agriculture 10,000 to 12,000 years ago was not the only period of time when dietary changes caused shifts in human genetic variation. As Beverly Strassman and her colleagues have concluded, "The forces of evolution (natural selection, gene flow, mutation, and drift) continue to act on human populations and have demonstrably altered allelic frequencies since the origins of agriculture. The best documentation of this is for malaria resistance and lactose intolerance" (Strassman and Duarte 1999).

Strassman further cautions against the view that our nutritional needs and optimal diet were "set" in the Paleolithic, for this view "ignores the fact that human evolution has been mosaic in form; different components of our biology evolved at different stages and rates. Our analysis of the transition to agriculture [from wild foraging] uncovered no empirical evidence that it was a singular watershed between adaptation and maladaptation."

The coevolutionary dance between our genes and our foods began long before the first farmers and herdsmen and continues to this day. The list of gene-food interactions in the introduction (see table 1) include genes that have been documented for some ethnic populations, but not others, making the notion of a single optimal diet for all of humankind an absurdity. The selection for these genes has occurred much more rapidly than the prophets of the Paleolithic prescription initially acknowledged. And that is why *evolutionary gastronomy* celebrates the influence of ethnic cuisines on both natural and cultural evolution *in our time*, as it does in the Paleolithic.

CHAPTER THREE

# *Finding a Bean for Your Genes and a Buffer Against Malaria*

FROM BALI and the Paleolithic, we move to Sardinia and the Neolithic, the era in which agriculture emerged as one more set of strategies for human land use and food getting. Although agriculture was once treated by archaeologists as a rapid revolution that stormed different continents at about the same time—some 10,000 to 12,000 years ago—it is now recognized that hunter-gatherers practiced elements of plant selection, transplanting, and dispersal for many thousand years before that apparently instant revolution. In other words, agriculture was a *slow* food revolution that seldom transformed any culture's traditional diet in one fell swoop. Neither was the adoption of agriculture a peculiar watershed between Paleolithic adaptation and Neolithic maladaptation. Farming and herding peoples like the Sardinians and Cretans continued to draw upon wild herbs, legumes, snails, and fish, and that is one of the reasons we will be visiting them.

There are, nevertheless, some gene-food-culture interactions that developed in the first few thousand years after the adoption of agriculture that have become easy to document through archaeology and written history. The best evidence of how these interactions developed comes from islands such as Sardinia and Crete, rather than from the continental cradles of agriculture. One such story from Sardinia tells how climate, disease, human land uses, and food choices all contributed to the selection of a certain gene that has two faces. One face makes it look like a genetic disorder, while another shows it to be a profoundly critical ecological adaptation. On Sardinia, it becomes clear that human gene/food interactions were not set in stone during the Paleolithic, with little change since; the Sardinians have experienced strong selection pressures from disease and diet that have uniquely directed this people's recent evolution.

Springtime in Sardinia is a good time for the birds and the bees, for the broad beans known as favas, for their flowers, and even for their pollen. It is a good time and place for contemplating food history as well, especially as we reflect upon how a particular agricultural diet such as that of the Sardinians has shaped human genetics differently than earlier hunter-gatherer diets had done.

Arriving on Sardinia's western shores during the height of spring, we see crops that sometimes run clear to the salt-sprayed cliffs. And yet, we can plainly see that this island has not been cultivated so long and hard that its wild plants and animals have been completely marginalized. Indeed, the coastal plains are ablaze with spring wildflowers of all shapes and colors, from white poppies and golden tossiliggines to spikes of deep, azure-hued squills, ballerina orchids, powdery blue gentians, and nodding cyclamens.

The Sardinian songbirds are nearly as conspicuous and colorful—singing from roosts in the hedgerows and wetlands that edge the agricultural expanses of the Campidano Oristano, the coastal flats of west-central Sardinia. Several of these songbirds will have arrived just the week before, after flying thousands of miles across the Sahara and then the Mediterranean; here, near the port town of Oristano, they have found enough nectar and insects to set up territories, mate, and nest. Yes, the spectacular abundance of the natural world still seems woven into the fabric of Sardinian life, especially in spring; this is the season when the reproductive urge seems to run rampant in the wild.

But as we know from some reading we have done before we arrive, a lot of Sardinians do not become so enthused or aroused by the coming of spring. Every spring season for millennia, residents have braced themselves for the warm season that brings with it the threat of malaria and, for some, physiological discomfort resulting from the interplay between diet and genes. It is that interplay that has brought us to Sardinia during this particular season, when its effects can be observed even by the untrained eye.

When in Oristano, we are invited into a Sardinian high-school classroom, where we confirm what others have told us: the teenagers there do not seem to be achieving the same heightened sexual energy as the flowering plants and courting birds just outside their windows. We can sense that a certain malaise had set in, as if the students had been drained of all their energy rather than enlivened by the vernal equinox. When I chat with a few of the students later in the day, after they have left school, many of them—especially the boys—complain of feeling sleepy, dizzy, or on the verge of vomiting.

From what we later learn by talking with doctors, springtime is habitually a tough time for Sardinian youth. It has been documented that disrupted sleep and nightmares plague many boys during this season. An emergency room might admit a teenager in the middle of the night who had been frightened when he got up to relieve himself and found his urine dark and bloody. Distressed, he had his parents take him to the local hospital, where he received blood transfusions until the color of his urine returned to normal.

We have arrived during the Lenten period, when many Italian Catholics fast from meat and dairy products, so it is tempting to attribute some of the seasonal malaise and maladies found among Sardinian youth to their religious fasting. But Catholic fasting at Lent does not explain why the youth of other Mediterranean ethnicities display some of the very same symptoms. Coptic Christian farmers living along the Nile suffer from many of the same seasonal difficulties even though their dietary restrictions during the arduous fast of El Soum el Kibir are somewhat different. More remarkably, Muslim youth living along the Tigris River—the ancient cradle of Old World agriculture—suffer from the same trouble, which has been known for centuries as Baghdad Fever, with symptoms much like well-known forms of anemia.

While there have been descriptions of this sickness in Greek and Persian commentaries for several millennia, it was only five decades ago that scientists realized that while Baghdad Fever is triggered by exposure to a variety of substances, it is an inherited susceptibility. When seen through evolutionary and cultural lenses, the malady is clearly the result of a genetic adaptation to certain environmental conditions and culinary traditions associated with irrigated agriculture in the Mediterranean basin.

Rather than Sardinian, Albanian, Greek, Egyptian, and Persian youth becoming vulnerable from fasting too long—that is, suffering from what they had *failed* to eat—it was what they *did* eat and breathe that triggered a genetic response in them. The trigger is eating fava beans or inhaling fava pollen. By doing so, these youth are fortified to fend off the most serious threat to their Sardinian ancestors: malaria, the number one killer of farmers on Mediterranean coastal plains over millennia. Although the side-effects of sleepiness, dizziness, nausea, and discolored urine weaken sufferers of so-called Baghdad Fever, resistance to malaria is actually enhanced.

Malarial epidemics have dramatically shaped and reshaped human populations in the Mediterranean over at least the last 5,000 years. As long as farmers have cultivated and irrigated coastal plains, they have been plagued by blood-sucking females of the anopheles mosquito. When wetlands- and ditch-loving mosquitoes bit into human flesh, they injected the microscopic *Plasmodium falciparum* parasites into human bloodstreams. These malarial parasites readily collect in the victim's liver, and after two weeks of infection, they burst the liver cells open and let their progeny out to invade more red blood cells. Roughly 3 billion people live today in malarial regions—that's half of all humans presently breathing. While most have access to drugs, hospitals, and mosquito control programs, there remain 800 million cases of malaria each year, resulting in an annual average of 2 million deaths. Of course, these recent levels of mortality are low compared to those suffered prior to effective strategies for mosquito control. Imagine how much higher death rates must have been in the Mediterranean before modern medical treatments and pest control strategies were prevalent. But we must also now try to imagine how many millions more would have died if this peculiar interaction between genes, fava beans, and malaria had not emerged.

Despite their partial resistance to malaria, most islanders in the Mediterranean region are not quite sure whether their strong reaction to favas is really a blessing or a curse. A Greek farmer on the island of Kéa explained to me his feelings of ambivalence:

> Some people have this problem: they cannot eat the beans, inhale the pollen, or even grill game birds that have eaten beans in the fields. This is what we call *favismo*. But the interesting thing is that some people who suffer from this sickness are the same who don't get malaria so bad. They are vulnerable in other ways, however: they can't take certain medicines and they are unable to smell naphthalene without getting sick. Once the doctors do an examination, the children who have this problem wear a tag on them at all times, explaining what they must not be exposed to.

~ By some simple twist of fate, just as malaria-carrying mosquitoes were finally being controlled by spraying DDT all across Sardinia, scientists in other parts of the world were revealing some startling global patterns:

1. Some ethnic populations have a reduced genetic capacity to produce certain enzymes and are therefore disabled when exposed to certain drugs or foods, but are protected from malarial parasites at the same time.
2. More specifically, 7 percent of all humans—around 400 million people—have genetic adaptations that help them resist malaria. For at least 100 million of these people, their consumption of certain staple foods and herbs changes the severity of response to this infectious disease, keeping the malarial parasite from maturing and reinfecting other cells.
3. Finally, no less than seventy-eight biochemical variants or alleles have been found to occur on the G6PD gene that can confer partial resistance to the *Plasmodium falciparum* strains carried by Mediterranean mosquitoes. This human genetic variation offers one of the best examples of natural selection working on our species' ethnic populations through differential exposure to certain infectious diseases and traditional foods.

Oddly, the breaking of the genetic code conferring Baghdad Fever did not happen in the areas where the illness was historically most severe: neither along the Tigris and Euphrates, nor along the Nile, nor even on the coastal plains of Sardinia. The motivating force behind the initial discoveries of malarial resistance was not an attempt to understand food interactions with genes, or for that matter, to understand natural selection. Instead, the cracking of the code began in the late 1940s when an Oxford graduate student of Kenyan background, Anthony Allison, proposed that the distribution of sickle-cell anemia in Africa might be linked to the prevalence of malaria there.

When Allison tested the blood of residents whose families had lived in Africa's malaria-infested areas for centuries, he found a high frequency of the paradoxical genetic condition we now call sickle-cell anemia. While some individuals died young, others were far less likely to be debilitated by the malarial parasite. The sickling mutation caused cells to collapse in capillaries where they unload oxygen, keeping the parasites from reproducing and spreading to other cells. When an individual is heterozygous—that is, carrying only one copy of the sickling gene—the red blood cells collapse just enough to kill the parasite but not the human carrier. The gene that killed some protected others from malaria.

About the time that the significance of this discovery was sinking in among American scientists, the U.S. government initiated malaria-related research at the Stateville Penitentiary in Illinois, for a completely different reason: the army was trying to understand why a number of African American and Italian American soldiers given antimalarial drugs in Korea were dying not from malaria, but from the drugs themselves.

In the early 1950s, most American soldiers sent off to fight the Korean War were routinely given a relatively new drug, primaquine, as a means to protect them from their likely exposure to malaria-carrying mosquitoes in the coastal wetlands of the Far East. But the drug that the Surgeon General had hoped to be a lifesaver was somehow responsible for killing a small but significant number of African American soldiers. At the same time, it triggered acute hemolytic anemia in 10 to 15 percent of U.S. soldiers of North African, Middle Eastern, or Mediterranean descent. The Surgeon General became so disturbed by this loss of otherwise healthy soldiers that he ordered the use of inmates in the Stateville Penitentiary "as volunteers" to allow the Army Malaria Project to determine the cause of these deaths.

And so, a series of experiments was undertaken that would hardly be ethically sanctioned anywhere today for the risk of death participants were exposed to. African American inmates were first screened for primaquine sensitivity, and then their blood samples were exchanged with those of "healthy" (nonsensitive) inmates through transfusions marked with radioactive chromium. When red blood cells from insensitive patients were injected into primaquine-sensitive inmates, these cells continued to be healthy and unaffected even when primaquine was administered to the inmates. However, when radioactively labeled cells from primaquine-sensitive inmates were given to otherwise healthy inmates to whom primaquine was administered, these red blood cells were immediately destroyed.

Soon, a young physician was called in from the University of Chicago to see if he could explain what had happened to the red blood cells. That physician, Paul Carson, appeared to be just the kind of biomedical sleuth the army needed, one with a penchant for elegant experiments that elucidated physiological responses to various chemical agents. The other Army Malaria Project researchers already suspected that the primaquine-sensitive inmates must be suffering from some kind of defect, genetic or otherwise, in their red blood cells. They hoped that Carson's ongoing research interests regarding an enzyme named GSH reductase might allow him to solve their problem, since GSH itself is essential to the integrity of red blood cells.

And so, Carson designed a series of ingenious experiments whose results he tersely reported in barely a page of *Science* magazine text in 1956. Carson confirmed that the blood cells of four of the primaquine-sensitive African Americans were genetically deficient in the enzyme GSH reductase. And it just so happened that this enzyme was absolutely critical to facilitating one of the three steps necessary for the direct oxidation of glucose into fructose in red blood cells. Exposure to primaquine triggered hemolytic anemia in those who had inherited this enzyme deficiency, just as it "starved" the malarial parasite of the oxidation reactions it required to grow and reproduce.

The readers of *Science* could not have realized it at the time, but Carson's report was among the first to ever document human genetic variation in response to a drug as a result of an inherited enzyme deficiency. In other words, the public was aware that certain individuals could inherit their susceptibilities to penicillin molds and fava beans, but it was not yet aware that members of diverse ethnic populations might suffer severe reactions to such drugs and foods as a result of their shared evolutionary history. Instead of suffering from allergies (such as those afflicting individuals debilitated by penicillin injections), these people have a reduced capacity to produce an enzyme needed for normal oxidation activity when a set of plant-derived chemicals (including primaquine) enters their bloodstreams. The particular enzyme with reduced activity—glucose-6-phosphate dehydrogenase—now goes by the nickname G6PD.

Over the last half century, G6PD has not exactly become a household catchphrase, even in some million households where lives have been saved due to the belated scientific understanding of its importance. Nor have we made heroes out of biochemical sleuths such as Carson or another geneticist who soon followed in his footsteps, Arno Motulsky. Professor Motulsky was just beginning his own biomedical career when he read of Carson's experiments; he was so inspired by them that he has spent the last five decades pursuing their implications for a variety of other diseases. His own trailblazing research has led to the founding of two additional subdisciplines of biomedical research, *pharmacogenetics* and *nutritional eco-genetics*.

Not long after his first reading of Carson's work, Motulsky recalled a quirky speculation—described below—by the pioneering evolutionary biologist and agnostic J. B. S. Haldane, who had published an essay in 1949 entitled, "Disease and Evolution." Perhaps it was placing Carson's experimental evidence in the context of Haldane's sweeping evolutionary theories that allowed Motulsky to accomplish such astonishing applications of both—applications that have extended if not saved the lives of tens of millions of humans on this planet.

Many biomedical researchers of the 1950s could get excited about the kind of cutting-edge lab science that Carson and his colleagues were doing. However, it seems that few besides Motulsky were as taken by the evolutionary logic that Haldane freshly applied to biological problem solving. At that time, evolutionary theory had not penetrated very far into the biological training of biomedical professionals. On top of that, Haldane's iconoclastic essays were not very user-friendly for conventional practitioners of Western medicine. In fact, when the likes of Carson and Motulsky were training to become scientists, they were more likely to hear of Haldane as England's leading atheist or for his politically motivated departure to work in India than for the relevance of his theories to medical practice.

This is the same Haldane who quipped that "the universe is not only queerer than we suppose, but queerer than we *can* suppose." Atheist-humorist that he was, Haldane also spun out the tongue-in-cheek argument that if evolutionary processes were *not* responsible for the great diversity of life-forms on earth, then it must be because of the existence of a God with "an inordinate fondness for beetles." As you might surmise, Haldane's wit was too rarified for many family physicians and lab scientists of the 1950s to even fathom.

Nevertheless, Haldane became somewhat of a hero to a peculiar group of biomedical researchers who shared his unswerving conviction that evolutionary insights could explain nearly any global pattern of disease, infectious or otherwise. As early as 1938, he argued that there were likely to be genetic differences among factory workers in their vulnerabilities to environmental toxins, and understanding those differences in an evolutionary context could save lives. Unfortunately, the medical profession lacked the screening techniques to follow up on Haldane's suggestion for several more decades.

Rather than being thrown off by Haldane's theories, Motulsky was intrigued by the old curmudgeon's contention that infectious diseases had been a main agent of natural selection in humans over the last 5,000 years. In his 1949 "Disease and Evolution" manifesto, Haldane speculated that so-called red blood cell "disorders"—such as sickle-cell anemia and thalassemia—were somehow adaptive and had likely evolved in response to chronic exposure to malaria, affording some protection from this infectious disease.

For his part, Motulsky recognized that the G6PD deficiency that Carson had discovered was just the kind of red blood cell "disorder" that Haldane had meant. After all, when an antimalarial drug interacts with this genetic deficiency, malarial parasites cannot continue to infect their host, for the drug has disrupted the oxidation reactions they require for growth, reproduction, and the spread to other cells.

Within a year of reading Carson's paper in *Science*, young Motulsky noted that researchers in Israel and Italy had put in place another piece of the puzzle: the very same G6PD deficiency that caused sensitivity to primaquine also caused similar reactions in people of Mediterranean descent who either ate green fava beans or inhaled excessive quantities of pollen from fava bean plants. Motulsky hypothesized that those certain people had their ancestry in places where malaria was endemic. He and his colleagues quickly devised a rapid screening technique to test this hypothesis, identifying individuals who suffered from G6PD deficiency and then determining by interviews if they had ever suffered difficulties from growing or eating fava beans.

Just two years after Carson's discoveries at the Stateville Penitentiary, Motulsky traveled to Greece and to Sardinia to see if his rapid-screening technique could verify that entire ethnic populations in malaria-infested areas experienced genetic interactions between malarial parasites and fava beans. As a control, he also screened Alaskan natives in mosquito-infested but malaria- and fava-free tundra landscapes where bean-farming was but a recent introduction.

Through these efforts and those of his collaborators in other countries, Motulsky confirmed that the distribution of favism alleles of the G6PD gene shadowed the distribution of malarial parasites and anopheles mosquitoes in the Mediterranean basin. As Motulsky began to rough out a map of where favism alleles had been recorded, they mirrored the maps of the *falciparum* microbe and a certain set of mosquitoes that carry it. In short, Haldane's speculation was right on the money: favism was not simply a "disorder" or "maladaptation" that handicapped males who were *hemizygous* recessives for this trait.

We now know that favism is a sex-linked condition, located on the X chromosome. Women who have two X chromosomes can have normal G6PD enzyme activity and no anemia if they are homozygous or heterozygous. At the same time, women with two copies of the recessive gene and hemizygous men with one copy have reduced enzyme activity and are somewhat anemic. However, it is hemizygous men—with no dominant or "normal" gene for enzyme activity at all—that are unusually deficient in enzyme activity, particular if they have the recessive G6PD Mediterranean allele. These men have a marked susceptibility to the disruption of oxidation reactions, which gives them an edge in resisting the spread of malarial infection but makes them unusually vulnerable to favas and drugs. The boys at the Sardinian school who dramatically responded to inhaling pollen—the same who had dark urine and a plethora of feverish symptoms around vernal equinox—are no doubt hemizygous recessives, receiving but one copy of the gene from a single parent.

When I recently spoke with Motulsky in his hometown of Seattle, where he is still active in research on human genetic responses to diseases, he reminded me that favism remains the most widely recognized example of genetic *interactions* with foods and drugs: "Favism is perhaps the best example of the very concept of pharmacogenetics. If you happen to have the genetic deficiency, it is not necessarily a problem in and of itself. If you eat fava beans but you don't express the deficiency—as is true with many Sardinian women—again, no problem. An antimalarial drug alone—in the absence of G6PD deficiency—no problem again. But the *interaction* of these three factors can be deadly!"

As researchers around the world began to notice other adverse genetic interactions to the many new drugs hitting pharmacies in the 1950s, the study of pharmacogenetics boomed. Motulsky's own research demonstrated that the hereditary bases of these abnormal reactions were different from that of immunological responses generated by toxic allergens.

Unlike many of the new breed of pharmacogeneticists, Motulsky did not restrict his interests to the human genetic responses provoked by newly developed drugs and environmental contaminants collectively referred to as *xenobiotics*, that is, chemicals foreign to our bodies; instead he tried to figure out the evolutionary underpinnings of these responses. After all, while primaquine and other drugs were novel environmental triggers to a genetically hardwired response that could escalate into hemolytic anemia, the fava bean was a long-standing element of the Mediterranean diet. It could not so easily be labeled a xenobiotic, that is, a substance foreign to Sardinians, Persians, or Egyptians, for these people had apparently eaten fava beans for millennia. As Motulsky recalled to me how he broadened his focus, he was reminded that looking at all substances that triggered genetically conditioned responses in humans as "novel contaminants" did not make sense: "I soon realized that pharmacogenetics was but one of several necessary inquiries that would have to draw on the same underlying theoretical [evolutionary] framework. And so I began to speak of nutritional eco-genetics, a term that my colleagues and I first used in print in 1974."

*Nutritional eco-genetics* has been defined several different ways over the years, but here is how I like to think of it. It is the interdisciplinary field that determines how the long-term consumption of certain sets of foods has historically shaped the distribution of human genetic variation. These foods have interacted within the genetic traits of certain ethnic populations to help these people deal with the prevailing stresses in their environment. Gradually, it has become clear that genetic polymorphisms have developed in response to deadly diseases such as malaria and to a variety of other factors driving biological and cultural evolution. And yet, at the time Motulsky and his colleagues introduced the term *nutritional eco-genetics*, the role of traditional foods in the adaptive radiation of humans into diverse ethnic populations was still far from settled. Settled it now is.

Keep in mind that well into the 1970s, few human geneticists were as convinced as Haldane had been that *natural selection*—by diseases or by other stresses—played a role as a driving force in the divergence of human populations. More and more medical students were learning the basics of genetics and molecular biology in a reductionist—almost mechanistic—sense, but few were well informed regarding the broader context that evolutionary biologists now call *ecological genetics* or even *population genetics*. That is, when doctors detected inherent physiological differences among their patients in responses to various diets, drugs, or contaminants, they referred to these differences as evidence of *biochemical individuality*, without asking whether any of that variation was occurring among populations with different ecological histories. Only a handful of epidemiologists and geneticists of that era had begun to think about human diseases, nutrition, and environment from a population-based evolutionary perspective, and the few who were interested split on whether random mutations or natural selection had accounted for the levels of human genetic variation detectable at the time.

In retrospect, it seems remarkable to most of today's evolutionary biologists that their predecessors would doubt that natural selection worked among human populations in response to diseases, diets, and environmental stresses. But as late as 1974, when some of the world's most famous biologists and anthropologists met in Austria to discuss the role of natural selection in human evolution, these scientists spent most of their effort considering whether random mutations could explain human variation better than could natural selection. It was as if they did not quite believe that environmental influences had continued to shape human evolution over the last ten millennia.

At the same time, because geneticists had recently begun to master new, rather complex biochemical and statistical tools, they had been spending less time grounding their findings in evolutionary theory. Tracking the long-term demographic influence of random mutations through genetic and statistical models was all the rage. Many geneticists had little remaining interest in testing whether their data best fit with the patterns of natural selection or of other evolutionary processes. And despite impressive presentations on how malaria was a natural selective force for both sickle-cell anemia and favism, scientists would only affirm at the end of the conference that "the clearest demonstration of the action of selection in a polymorphism in man involves the sickling gene," which is responsible for sickle-cell anemia among Africans (Salzano 1975). Oddly, they argued that "direct data on differential mortality [for malaria-exposed carriers of favism] are lacking," so they withheld their support for Motulsky's evidence that G6PD deficiency also protects against malaria.

Despite this setback, within another quarter century, nearly all malaria researchers had come on board the natural-selection bandwagon. Today, few would argue that the distribution of G6PD deficiency reflects natural selection through malaria. In fact, it has been definitively established that the areas in Sardinia that have the highest frequencies of G6PD-deficient individuals are exactly where fewer deaths from malaria historically occurred.

~ Why was Sardinia the first place that scientists amply documented recent natural selection via human responses to disease and diet? Perhaps it was because prehistoric and historic Sardinians had suffered more from the *falciparum* parasite than any other human population, for conditions on their island were so conducive to malaria-carrying mosquitoes. As Peter Brown succinctly put it, "For two millennia, Sardinia was the most malaria-stricken region of the Mediterranean. The disease was seasonal, hyperendemic, and the greatest single cause of mortality" (Brown 1986).

Malaria was undoubtedly present in Sardinia by 4000 BC—as analyses of prehistoric skeletons have confirmed—but when the island's population peaked at 350,000 residents at the time of Christ, it was ripe for the spread of this infectious disease. By 1300 AD, the island's human population had dwindled to only 80,000. Over the next 180 years, half of all Sardinian villages were abandoned as a result of the disease, but the losses were far worse near the coastal wetlands, where 72 percent of the farming villages disappeared. Although Sardinia's population began to recover during the nineteenth century, again, during a six-year period in the 1920s, a half-million Sardinians suffered from malaria. It was not until after World War II, when the Rockefeller Foundation paid for the spraying of DDT in every corner of the island that mosquitoes were eradicated to the point that the malarial parasites could no longer spread. At last, malaria was no longer the driving force in the natural selection of the Sardinian people.

The imprint of the *falciparum* parasite was indelibly written in Sardinian genes, even though the threat appeared to have declined for a while. Sardinians no longer had any need to expose themselves to primaquine, but a number of other exposures triggered hemolytic anemia in G6PD-deficient individuals, including the eating of fava beans. It seemed paradoxical to many scientists that if eating fava beans or even inhaling fava pollen triggered the same hemolytic anemia that drugs did, that Sardinians had not long ago selected the beans *out* of their diet.

And yet, go to any seaside *ristorante* in Sardinia, and you can order as an appetizer a plate full of semimashed favas, cooked with a half-dozen spices, then cooled and dowsed with lemon and garlic and placed in a pool of deep green olive oil. The same is true in Egypt, where they are called *ful mudammas*, or on Crete, where they are called *koukia*.

As mentioned earlier, it must be remembered that Sardinians and other dwellers of the Mediterranean basin have long expressed a pronounced ambivalence about these beans. A Greek historian put it this way, as we sat sipping anise-flavored raki in a sidewalk cafe in Iráklion: "They are a little bit of medicine, especially for malaria, but a little bit of poison, some say as well. But that is how it is for a lot of what the poor can afford as foods, no? Up through World War II, we had no choice but to eat them, we were so poor. We ate everything: green beans, dried beans, even the sproutlike tendrils and leaves of the plants."

This two-sided view of favas is deep-seated and balances a certain respect for their properties with an undeniable dread of both the anemia and malaria associated with them. The Greeks treated the beans themselves as if they carried the supernatural force that they called *lepos*, which embodied the souls of one's parents as well as the seeds for future reincarnation. In other words, favas were a traditional crop that embodied your own ancestry as well as your destiny, making it dangerous not just for eating but for sowing as well. As cultural historian Alfred Andrews described in 1949, fava beans were part of ancient rituals in Italy and Greece, being offered to deities at special ceremonies in late spring or early summer. Nevertheless, the Priest of Jupiter was forbidden to touch a fava bean or even say its name, and Pythagoras instructed his followers to abstain from eating them or even entering a field where the beans were planted. Both Pythagoras and Pliny lived in a time when it was commonly asserted that the souls of the dead are in fava beans, and with that in mind Pythagoras compared their consumption to human cannibalism: "It is an equal crime to eat [fava] beans and the heads of ones' parents."

These cautionary words, laden with a sense of terror and respect for a plant that both grew wild and in cultivation in southern Europe clearly echoed sentiments that predated classical Greek and Roman philosophers. Indeed, as Andrews observed, "no plant or animal known to Indo-Europeans produced a more luxuriant growth of beliefs than fava beans" (Andrews 1949). The ancient farming folk of the Mediterranean believed that fava beans chewed then exposed to the sun would come to smell of human blood or sperm and that green unripe beans left in a pot would spontaneously turn into blood. While some believed the beans to be an aphrodisiac, others thought that they caused so much bloating and flatulence that their gasses could transform things before your very eyes. It was commonly believed that even wilted fava bean blossoms captured inside a vessel would produce so much powerful gas that they would change into the head of a child or a woman's pudenda. Favas most ancient name in Greek is derived from an even older Indo-European cognate for "blood" or "bloodiness."

Nearly all Mediterranean peoples have continued to use fava beans as staple foods, especially in the spring. The most susceptible members of their populations are cautioned from eating green beans and matured beans with the seed coats still on them. And yet, a significant portion of these populations remain vulnerable to favas. This paradox has intrigued a different sort of scientific sleuth, nutritional anthropologist Solomon Katz, who since 1976 has demonstrated the many ways the cultural histories of fava beans, culinary practices, and human genes in Sardinia and other lowland Mediterranean locales are inextricably linked. Picking up where Motulsky left off, Katz noticed that nearly all G6PD-deficient populations of humans appeared to have had fava beans as a major component of their springtime cuisines for centuries if not millennia. The timing of fava bean harvesting and consumption on each island or in each country around the Mediterranean, Katz observed, coincides with the beginning of the mosquito season and the upswing in exposure to malaria.

Katz has asked two intriguing questions. Could fava bean consumption by those exposed to malaria have triggered natural selection of those who are G6PD deficient? In other words, is fava bean consumption a cultural analog (in reducing oxygen available to malarial parasites) to what eventually emerged as a genetic adaptation among Mediterranean ethnicities? It was not lost on Katz that the ecogeographic distributions of wild and domesticated favas perfectly overlapped with the distributions Motulsky had found for the *falciparum* parasite, anopheles mosquitoes, and the Mediterranean favism gene for G6PD deficiency.

What Katz and his collaborators proposed was a model of *biocultural coevolution* to explain how fava bean consumption and G6PD came to protect Sardinians from malaria and how orally transmitted cultural knowledge and culinary practices tend to reduce the health risks associated with favism. One wintry day, I met Katz for dinner in his hometown of Philadelphia, where he teaches at the University of Pennsylvania; favas were not on the menu, so we shared a bottle of pinot grigio and a huge bowl of clams. As our conversation proceeded and the wine settled into our bloodstreams, I sensed that Katz had thought through the favism story in such depth that he could hardly get it out at one time.

"The coevolutionary process must have kicked in about the time the Greeks, Italians, and Sardinians established sedentary farming villages near the coastal wetlands in the Mediterranean basin—say, 5,000 years ago," he said. "They had eaten wild fava beans perhaps since their arrival in the region and had cultivated domesticated favas on a small scale for several millennia, but it was not until they were sedentary and adjacent to natural wetlands and irrigation tail waters that malaria began to be a strong selective pressure for this interaction."

I was surprised. "Wait a minute, Sol—in less than five thousand years? That's pretty rapid selection for the favism alleles that confer whole or partial resistance to malaria, isn't it? What, fewer than 200 to 250 generations of humans? What about all this stuff from your anthropological colleagues that says humans have remained essentially the same genetically since Paleolithic times?"

"Well, I'm pretty convinced that biocultural evolution can happen on this time scale—look at the development of lactose tolerance in herding cultures since the domestication of livestock. Now, there are some scientists that believe that the selection could have begun earlier than fava bean domestication, since there were many plants in the Paleolithic diet that have some of fava's properties, but the whole process must have accelerated dramatically when farmers began to live year-round on the coastal plains rather than escaping into the mountains during the malaria season."

Sol Katz had opened up a possibility that I had yet to consider—that human genetic adaptation to diseases like malaria and foods like favas could occur in a matter of a millennium or two. The time to fix such a gene (as G6PD) in an ethnic population depends, I surmised, on how potent the dietary chemical is and how lethal a disease is.

The properties of fava beans, I later learned are unique in some ways but not in others. Green immature beans and the seed coats of dried favas are rich in some powerful glycosides, as are New World lima beans. Upon ingestion, the glycosides in lima beans can be hydrolyzed into small but potent quantities of cyanide; in favas, they are hydrolyzed into divicine and isouramil, both of which function as pro-oxidants. While we are usually told that antioxidants are good for us because they bind free radicals that foster cancerous growth, pro-oxidants potentially increase the formation of those nasty free radicals. But these particular pro-oxidant compounds also undergo reduction-oxidation reactions until they deplete a compound known as GSH, which is essential for maintaining the integrity of red blood cells. Katz learned through some simple experiments that if you eat too many fava beans, your GSH levels plummet just as they do when you are given an antimalarial drug. The diminished GSH levels interfere with the growth and replication of *falciparum* parasites, thereby offering any fava bean eater temporary resistance to this infectious disease, but the resistance is clearly enhanced in carriers of a G6PD-deficiency allele.

Good sleuthing, but does this necessarily mean that carriers of favism and fava beans coevolved? Some critics have argued that Katz is too casual in his use of the term coevolution to describe the interplay between genes and orally transmitted culinary traditions associated with favas. Katz based his claim on the fact that Sardinians and other G6PD-deficient ethnicities of the Mediterranean know how to reduce the health risks of favas' divicine and isouramil content by removing seed coats or avoiding the beans when they are still green and unusually potent. By his logic, G6PD-deficient individuals would not have their malaria resistance seasonally enhanced if they did not learn from their elders how to properly prepare and eat favas during the season when they are most needed.

What is perhaps even more interesting about this traditional knowledge is not merely its ambivalence about favas themselves, but the way it may use herbs to push the valence in one direction or the other. As I will explain below, the herbs that are added to a cooking pot of slowly simmered favas can either be pro-oxidants or antioxidants; they can either potentiate or mute the effects of favas and G6PD deficiency.

I learned of this basic principle not in Sardinia but in Hawaii, talking with medical anthropologist Nina Etkin, who has studied antimalarial herbal treatments all around the world. Etkin now teaches at the University of Hawaii in Manoa Valley, where we met one summer morning for coffee and green tea, science talk, and gossip about mutual friends. As Etkin learned what I had been talking about with Katz, Motulsky, and others, she gently shifted my inquiries away from an exclusive focus on favas to consider other pro-oxidants as well, and the interactions among them: "Gary, take a look at the oxidant potentials of traditional medicines and foods that achieve virtually the same resistance against malaria without the high physiological costs of genetic predisposition."

At first, I did not realize the significance of the lesson that she was trying to teach me. But when she loaded me up with reprints of articles she had penned and books on malaria that she had contributed to, I took them and mused over them for months. And then one day, when I happened to be rereading one of her essays on plants as antimalarial drugs, I did a double-take at a section on herbs and spices that function as pro-oxidants. She listed, among others, rosemary, cinnamon, nutmeg, garlic, onion, basil and clove. Hadn't I just read some rather similar lists of herbs somewhere recently?

When I paused and realized where I had encountered those other, nearly identical lists of pro-oxidants, I had a eureka moment: the day before, I had been rummaging through Claudia Roden's classic, *The New Book of Middle Eastern Food*, Patience Gray's extraordinarily beautiful *Honey from a Weed*, and Clifford Wright's *A Mediterranean Feast*, searching for authentic recipes for fava beans. These were some of the same herbs used to spice favas for various occasions, while at other times, antioxidants such as chiles and oreganos are used. It was as though potherbs mixed into a fava cooking pot could accelerate or brake the GSH-depleting effects of isouramil, divicine, or the G6PD deficiency itself, depending on the mixture of herbs used.

These combinations, as I learned from reading Egypt-born-and-raised Claudia Roden, are not random concoctions; many have been blended together for so long that they are well known throughout the Mediterranean and codified by folk names widely used along the spice trails between Baghdad, Damascus, Beirut, Cairo, and the Casbah, or between Athens, Palermo, and Oristano.

*Baharat*, I soon learned from Roden, is typically an Egyptian mixture of ground cinnamon and cloves with allspice or rosebuds. It leans toward the pro-oxidants. *Ras el hanout*, the "grocer's head" mix blended by spice merchants in North Africa, contains cinnamon bark, nutmeg, cloves, ginger, and various peppers, sometimes with the aphrodisiac Spanish fly thrown in for good luck. It too is dominated by pro-oxidants. But *zaatar*, the blend based on wild mountain thyme that my Lebanese kin collect in the hills above the Bekáa Valley, includes salt, sesame, and ground sumac berries, probably resulting in a mix with more antioxidants than the others.

And so, the dominant spices thrown into the fava pot may vary from place to place but are fixed by ethnicity and/or season. The Catalan dish, *faves guisades*, "sweats" fava beans in an earthenware olla with rosemary, thyme, garlic, onion, oregano, and parsley. By drenching the beans in lemon juice, or by adding more of some herbs than others, you can either quench oxygen molecules or let them loose. And which way you push the redox reactions can either increase your resistance to malaria during the height of the season or dampen the fava side effects so that you don't go anemic.

It is not that the folklore associated with favas and herbs gives cooks a precise chemical formula; trial and error subliminally guides us in their use. The same pro-oxidants and antioxidants keep the plants themselves healthy, helping them to heal wounds, repel pathogens, and tolerate drought stress. Are the plants conscious of how their chemicals work against various stresses? No, but do these strategies work? Yes, and they have done so for hundreds of millennia.

Is it any surprise that the same plant chemicals have kept dwellers of Mediterranean coasts from suffering so terribly from malaria, even though their users did not know the exact chemistry in play? Few medicinal plant chemists remain surprised by such properties; they are nevertheless awed by just how efficaciously some of the plants help us.

But the larger issue—of how food plants' chemical arsenals have literally shaped us—is one that even geneticists are trembling about.

And to comprehend how deeply ethnic populations' very metabolisms have been shaped by the mainstays in their traditional diets, we now travel from Sardinia, around the Italian boot and its Sicilian kickball, until we land on Crete. It is there that the range of foods comprising one of the many healthful Mediterranean diets gives us a greater sense of the synergies within traditional cuisines.

CHAPTER FOUR

# *The Shaping and Shipping Away of Mediterranean Cuisines*

WE JOURNEY SOUTHWARD, then westward on this stretch of our odyssey through the islands of the Mediterranean. Out on one of the driest islands in the midst of those turquoise seas, we are searching for the nursery grounds for a rather strange bird, one known across the world today as Mediterranean cuisine. This traditional diet is supposed to fit all of us whose forefathers adopted cultivated foods and left the Stone Age diet behind. Although elements of the Mediterranean diet have migrated to nearly all corners of the earth, to know it fully, we must know it where it originally nested.

On the island of Crete, we can observe how a complete cuisine emerged as an adaptation to a particular land- and seascape. Relatively isolated from other sources of food, Cretan cuisine demonstrates how a culture constructed an integrated diet out of the unique nutritional resources that the land and the surrounding sea offer. There, among the descendants of ancient Cretan cultures, we can more deeply understand that true adaptation to place comes from something more than a single gene being linked to a particular ethnic food; rather, a consort of foods shape and are shaped by many genes and cultural behaviors.

As soon as we arrive in the harbor of Iráklion, we immediately indulge our taste buds in foods saturated with the native flavors. We purchase from market stalls a variety of foods grown on the island, from olive oil to sheep cheese to raki and ouzo. They are not hard to find, since so little produce found on Crete has been imported. Should we encounter some imported produce, it hardly compares in freshness, taste, or texture with that plucked from the stony, often salty or limey ground of this island, the southernmost reach of what is currently considered to be the European Mediterranean. Gertrude Stein's quirky epiphany could hardly apply to any relationship between land and food as much as it does to Cretan cuisine: "After all, anybody is as their land and sea and air is. . . . It is that which makes them and the arts they make and the work they do and the way they eat and the way they drink" (Stein 1990).

To be in Crete and to eat like a native, we have but two choices. Once we have left the harbor, we can either hug the shoreline, feasting in the Venetian ports and tourist resorts, or we can climb high into the mountains. There, we will find villages of farming folk perched precipitously above sheer cliffs, for every patch of earth that is more arable than the bedrock outcrop of the cliff face has been shaped into vineyards, olive orchards, bean fields, or else has been fenced as pasturage for sheep and goats.

The squid, mullet, bream, octopus, *ouzeris*, and the bouzouki music of the coast have their own enchantment, to be sure. But the highlands of Crete offer something altogether unique, easily distinguished from the rest of Greece and even from the coastal plains of Crete: a traditional cuisine that has changed very little over the last 6,000 years. At its core, Cretan cuisine still swims in olive oil, with a ballast of wild bitter greens, various beans such as garbanzos, large whites, and lentils, and barley rusk biscuits, all washed down with grapes fermented into retsina or distilled into ouzo and raki. There are also land snails, sometimes lamb, goat, shellfish, or anchovies in season. But olive oil, beans, and greens are woven into most meals—pull one away from the others and we will have unraveled the essential threads that make Cretan culinary traditions so distinctive.

Crete's cuisine also turns out to be reputed to confer to humans the greatest lifespan and the lowest rates of heart disease anywhere in the world. Within the last two decades, tens of millions of people from five continents have read books and magazine articles that tout Cretan culinary traditions as the model diet for long and healthful lives. Because some of these readers had already survived heart attacks, they have been searching for an antidote to further suffering. Many of them have been convinced by their physicians that religiously following a Mediterranean diet is their last best hope for remaining amidst their loved ones awhile longer. Although good survey numbers are lacking, it is likely that several million people on the planet at this very moment are attempting in their own peculiar way to eat like a Cretan—even though the majority of them have never been to Crete.

While there may be no such thing as a one-size-fits-all diet that can stretch to serve the bulk of humankind, many food writers have nevertheless argued that for dwellers of urban and agricultural landscapes, the Mediterranean cuisine is an easy one to adopt, regardless of whether your ancestors come from Crete.

And that is why my wife Laurie Monti, a medical anthropologist, has come along as part of a small group on a week-long visit to this island: to see if a seasoned public-health practitioner thinks that this particular culinary legacy can provide a healthy diet for most contemporary Europeans and European Americans. Laurie has worked with traditional diet, medicine, and health concepts in a half-dozen countries around the world, but after serving as a nurse-practitioner with many at-risk populations, she also understands what Western medicine has to offer in terms of accurate health assessments.

Questions about the adoptability of the Mediterranean diet for health benefits are not trivial, since this diet is now being used to target the number one cause of death in the Western world: heart disease. When heart disease is combined with arteriosclerosis, diabetes, cancers, and other chronic illnesses, such diseases of malnutrition account for well over 60 percent of all deaths in the United States. While there are other traditional cuisines found around the world that may also reduce the impact of these diseases, it is the Mediterranean diet that has received the most play.

Geographer Leland Allbaugh was perhaps the first to argue, as he did in 1953, that the Cretan diet in particular has remained in step with the nutritional needs of active (rural) people. As has been recounted often in recent years, the follow-up test of Allbaugh's hypothesis was the Seven Countries Study coordinated by Ancel Keys of "K-Ration" fame. The Seven Countries Study was one of the first comparative health surveys of its kind that was anthropologically informed as well as quantitatively accurate in its nutritional assessments. Keys and his colleagues determined that highland Cretans suffered a coronary death rate of 9 persons per 100,000, far lower than any other population in the study. Surprisingly, the population of Americans sampled had coronary death rate forty times higher, even though the Cretans were consuming almost three times the fat that the Americans were and one and a half times more than other Mediterranean populations studied.

The team of epidemiologists that accomplished the initial research in Crete must have worried that it was presenting a heresy—that more fat was better—but a 1987 World Health Organization study soon bore out their results. The WHO found only 7 coronary deaths in Crete per 100,000, with Americans still suffering thirty-seven times higher death rates due to their diet and exercise regimes. Of course, the team quickly clarified that it was not simply endorsing fat consumption, but that there was something about the particular use of olive oil in complement with other Cretan foods that merited further attention.

The original profile of Cretan health was undertaken in the mountain village of Spili, so with Laurie and me in the lead, our group headed inland and upland, escaping the blistering heat, clutter, and traffic of the beach towns. Our little rental cars rose up, up, up along winding roads that meandered through maquis scrub pastures for goats and sheep, and past vineyards where many of the grapes are grown with trellising. We passed a road that leads to a vineyard just over the mountain from Spili that has been in continuous cultivation for no less than 2,500 years. Nearby, the same stone press has been used for mashing grapes for more than fifty generations. This is not exactly the Land of Planned Obsolescence to which Americans have grown accustomed.

As we climbed in elevation, we left the open landscape of vineyards and vegetable fields and crossed ravines harboring dense groves of walnut, carob, and chestnut, sometimes with vine crops planted beneath them on ancient terraces of dry-laid stone. Thousand of kilometers of stone walls hold the soil in check, and we could palpably feel that the land had been worked by human hands. It had been that way for more than 7,000 years, when Neolithic colonists first began to manage wild olives and hunt the feral goats known as *kri-kri*.

Coming into Spili just before dusk, it seemed to be shrouded in even richer, deeper greens: thick-trunked cypress, pine, and fig trees lining the roadside, gangly stalks of artichokes up against fences, rampant grapevines spilling over the porches and patios of homes and cafes, aromatic herbs and gardenias shading stairways and walkways, and most impressively, dooryard gardens covering every square inch of open ground. Spili is nestled below the limestone cliffs and bulbous ridges of Mount Kedros, where giant griffin buzzards still soar. A cool breeze came up as the sun went down, and soon we watched the full moon appear. It did not "come up" so much as it emerged horizontally, from behind the cliffs of Kedros. We parked and walked along the six ramshackle blocks of downtown Spili, looking for a *taberna* in which to enjoy our first meal of mountain cooking.

Almost immediately, we noticed the large number of elderly present among the community of Spili: Orthodox priests with long gray beards; widows who have worn black dresses since their husbands were killed during World War II; great-grandmothers teaching young wives how to weave lace; balding farmers whipping the butts of mules that carry loads of forage to backyard chicken coops. If there is a place in the world where elders are amply present to be able to pass on ancient food traditions to younger generations, it is here in the shadows of Mount Kedros.

Then again, there is a touch of self-consciousness among the restaurants in Spili these days. Although the first sign we passed on our way into town announced "Fast Food: Suvlaki, Gyros," many of the other tabernas claim to serve "authentic Cretan cuisine." Books on "Cretan cooking" and the "Mediterranean diet" fill the shelves of gift shops, and boutiques sell local olive oil, dried herbs, and thyme honey in attractive jars and baskets. After all, Spili is what scientists would call "the type locality" of the Mediterranean diet, where its health benefits were first described with unremitting accuracy.

Of course, most of us who had come this far already realized that the Mediterranean diet is not "one species," but a mosaic of variation from Spain through Sardinia, southern France, Italy, Sicily, Corfu, Greece, and Asia Minor. Nevertheless, most nutritional studies from the region, as well as dozens of Mediterranean cookbooks appearing since the late 1970s, all pay homage in one way or another to the original studies done in Spili not long after World War II. If any village deserves to capitalize on its culinary legacy, Spili does.

~ In 1948, as the Greek government tried to reconstruct its country after the war and to deal with the devastating poverty of its rural areas, it invited the Rockefeller Foundation to undertake epidemiological studies that could suggest the most effective means of improving the health status and standard of living of its citizens on Crete. That was when the American team of epidemiologists led by Leland Allbaugh undertook the first systematic studies of diet and health on the island, interviewing one out of every 150 inhabitants about their eating patterns.

Allbaugh and his colleagues reported that 61 percent of the calories eaten by Cretans each day came from fruits, vegetables, greens, nuts, and roots, almost twice the amount of plant intake for the average American at that time. Even more curious was the fact that Cretans were eating more fat than Americans, but that 78 percent of the fats served on the Cretan table were the monounsaturated ones of olives and olive oils. Despite their poverty, Allbaugh had to conclude that Cretans were eating a healthful diet, *one that he supposed had remained remarkably unchanged for forty centuries*.

In truth, during those forty centuries, as various invaders came and went, they brought along a few handfuls of rice from the Far East, potatoes and tomatoes from South America, zucchinis and green beans from North America, and, most certainly, coffee and tea. And over the last two centuries, certain tree crops such as carob, chestnut, and acorn-bearing oaks have been neglected for their food products and have been used more for their wood and shade. But a number of factors have combined to ensure that the Cretans maintain many elements of their ancestral diet, even as many other ethnic populations around the world have lost theirs. While there has been incremental change in the diet across many centuries, Crete has not suffered the wholesale loss of its traditional diet as has happened in so many other places.

Residents have not simply kept many traditional foods in their gardens and on their plates. They have somehow retained the traditional knowledge of how to seasonally seek out and prepare the wonderful range of wild and managed foods placed before us on the tables of the *tabernas* and *ouzeris* of Spili.

"These snails I'm serving you," one chef proudly declared, "I personally harvested them from a grove above the village that the creek runs through. I begin looking for them under bushes in March, when the ground is still wet in the mornings and all the wild greens are coming up. I take them home alive to put in a special box, then feed them rosemary and flour until they lose their bitterness. It is then that they are ready to serve my visitors."

With that announcement, he set before me a bowl containing three or four dozen tender-meated terrestrial snails, glistening with oil droplets and flecked with rosemary. They were floating in a pool of extra virgin olive oil, and they were delicious.

As was nearly everything we ate in Spili that week: delicious and dowsed with olive oil. The simple Greek salad—*horiatiki salata*—was also floating on a pool of olive oil, as were the boiled amaranth greens, the sage-laden lamb chops, the rabbit legs baked in clay pots, and the green beans in tomato, lemon, and oregano. By the third day, my gut microbes asked for disaster relief because my GI tract had been hit by an oil spill—I was suffering from stomach cramps simply because my fat consumption had tripled in a matter of days. Others among our group suffered the same discomfort.

I was a chagrined that we could not quickly charge ahead in our efforts to eat like true countrymen of Crete, for the early studies of their traditional diet all touted olive oil as the key factor in conferring health benefits and longevity. The regular daily consumption by elderly Greeks of 50 grams or more of virgin olive oil was shown to lower their triglyceride levels and to improve HDL / LDL ("good" cholesterol/ "bad" cholesterol) ratios, and both factors reduced their risk of heart disease. What's more, as the only edible oil from a crop plant that contains polyphenols, olive oil is endowed with an artillery of powerful antioxidants that fight cancer. As these discoveries reached the popular press, the demand for Greek olive oil exploded, and soon young Cretans were bulldozing open new orchards to increase their family's production of this liquid gold.

Talk to any native of the Cretan countryside and you sooner or later learn that they need not be convinced by scientists that olive oil should be a favorite fat. After their faith in Jesus, Mary, and the menagerie of Orthodox saints, Cretans believe in olive oil. It is not merely used for dressing salads and cooking vegetables; it also serves to marinate and baste meats, to cure and preserve, to drip and drizzle. Aside from its many culinary values, Cretans also use the essence of olives in lamps, soaps and shampoos, facial creams and foot powders, as a medicine to heal wounds, and as an ointment with which to baptize babes and to bury the dead. The olive has been considered the sacred tree of the island since the era of the Minoans. About the only use of olive oil that Cretans know well but have not developed a retail market for is as an elixir to ensure conception, owing to its aphrodisiac qualities. From the viewpoint of a Cretan merchant, since Adam and Eve undoubtedly discovered this aphrodisiac during their stay on the island long ago, it would be presumptuous of any single Cretan to patent a product based on that ancient use.

The fact remained that after several days in the highlands, I was drowning in olive oil, so I had to reduce my consumption of it for several days. I later mentioned my inability to immediately accommodate large doses of olive oil to Dr. Antonis Kafatos, a pediatrician and nutrition researcher at the University of Crete who has been pivotal to studies of the Cretan diet for more than three decades. A handsome, fit, white-haired native of Crete trained at Columbia University, Kafatos was not surprised by my condition.

"When we've begun to serve the same olive-oil-based meals to Cretans and to British and Irish in our dietary studies, there is an immediate metabolic difference apparent. We find faster postprandial [after-meal] clearance of blood lipids in the Cretans. But I can't say that it is a genetic difference between the two populations, because within three to four weeks on the Cretan diet, the clearance rates are approximately the same."

While Kafatos and his colleagues were successful in getting the British to increase their consumption of olive products, a research team working in Lyon, France, had more difficulty in getting its coronary patients to significantly increase their olive oil consumption. The French do not necessarily eat "Mediterranean," in the sense that olive oil comprises less than a tenth of their total consumption of vegetable oil, while margarine and butter make up more of their total fat intake. Given their penchant for butter, the Lyon heart diet study team complained that they had to hook patients on a canola oil margarine resembling olive oil in its health benefits: "In duplicating the Cretan diet, we confronted the problem that it is impossible in practice to impose olive oil as the only edible fat on populations unfamiliar with its taste" (Renaud et al. 1995). But it was not just olive oil that the French could not stomach. Even though some three hundred French patients were specifically trained and actively encouraged to "eat Cretan," they ended up eating only a fifth of the oil, half the fruit, three fifths the legumes, and half the bread that Cretans traditionally ate. What's more, they continued to drink ten times the alcohol, eat three times the fish, and a third more meat than the Cretans.

Despite their mixed performance in approaching an authentically Cretan dietary regime, the French heart disease patients who persisted in trying to shift their eating patterns toward that ideal suffered 70 percent fewer repeat heart attacks and lived longer than the control group. But the experiment simply begged the larger question: is the authentic Cretan diet and its health benefits fully transferable to any other population living away from Crete?

A Greek research team hit pay dirt when designing another study. In selecting a group of people more inclined to eat Cretan than the French, they could not have picked a better group: a community of Greek immigrants in arid coastal Australia who retained a fondness for their ethnic culinary heritage. But, again, the results were mixed.

When I met with one of the coauthors of the Australian study, I asked what I thought was an obvious question: is every Greek indeed adapted to consuming the 31 kilos of olive oil ingested annually by traditional Cretans? There was a pause, a lingering silence.

"Well, we don't yet know. Well, yes, we know that there are health benefits when certain high-risk groups approach the Cretan diet, but we have not yet studied human genetic variation in response to that diet."

I later learned of a comparison of northern Europeans placed on the same diet as residents of Crete. Tellingly, the triglyceride and hormone responses of the islanders were less affected by fat consumption than were those of their northern European counterparts. For thirty young males from Crete, blood lipid levels after meals showed rapid returns to fasting concentrations of triglycerides and apolipoprotein B, thereby reducing their risk of heart disease. Curiously, Cretans had healthier responses in blood coagulation activity on a high olive oil diet than they did on a diet of saturated fats, and on both kinds of fats, their responses were noticeably different than those of northern Europeans. This study by Zampelas, Kafatos, and their colleagues should have come as no surprise to any medical researcher who has been following the growing literature on apolipoproteins B and E in the context of gene-diet interactions.

Carriers of different lipoprotein alleles have markedly different responses to high-fat diets, with the carriers of some alleles suffering from elevated cholesterol levels while carriers of other alleles show negligible effects. In addition, the *kind* of fat or oil produces different responses among different ethnic populations. In short, all consumers of olive-based *to eleoladho* and other edible oils are not created equal. After centuries of consuming the largest quantities of olive oil of any people in the world, Cretans have evidently developed a genetic adaptation to these levels of consumption. If you are from some other ethnic ancestry, you can come to Spili to eat like a peasant, but your body will not absorb the Mediterranean cuisine just as the body of a Cretan would do. It's as simple as that.

~ Dr. Kafatos also reminded me that there is something deeply cultural, not just genetic, about the way Cretans eat—and about when they do not eat—that not everyone in other populations can emulate. Frankly, his charm could convince Laurie and me of this, even if all the data were not yet in. As we sat in the old port of Iráklion, overlooking the sea, he talked about the flip side of eating—fasting—which he thinks has as much to do with health as feasting does on Crete.

"We've been tracking 120 Cretans," he explained to us, "sixty of whom strictly observe Greek Orthodox religious fasts that occur over 180 days each year."

"They fast for half the entire year?" I asked, astonished. I knew that my Greek friends fasted for Lent and prior to Christmas, as well as prior to the Assumption of the Virgin Mary, but I had casually guessed that these observances amounted to ninety days per year.

"No it is twice that number of days," Kafatos replied. "Of course, individuals vary in the severity of their observance. Most forego meat from livestock and dairy and eggs, while others forego olive oil as well. For many, land snails, calamari, and other invertebrates are acceptable. There are days, however, when one must eat fish or not eat fish."

"What are the effects on the health of the fasters?" Laurie asked Kafatos.

"It's significant—about a 12 percent reduction in their serum lipoproteins—the amount of fat their blood flow is carrying around. It has a measurable effect on their body fat composition as well."

"I can't imagine most Americans would tolerate fasting for even a week a year," I replied. "I happened to fast for eighteen days over Lent this last year, and some of my colleagues treated me as though I were attempting suicide."

Kafatos chuckled, and then turned serious once more. "It's not just the abstinence from certain foods—we also take special advantage of the seasonal wild plants that are available during these periods of religious obligation. I'm sure you've talked to Nikos and Maria about the many *aghria horta* that appear right around Easter."

Kafatos was referring to Crete's acknowledged experts on wild bitter greens, Nikos and Maria Psilakis, scholars and journalists who have championed the maintenance and revival of traditional foods on Crete. Indeed, Nikos and Maria had brought us a bouquet of greens and herbs from their own garden just the night before. As Kafatos implied and as the Psilakis have documented, Cretans continue to use at least 150 wild plants for food over the course of a year—with a different group of greens each season. Coincidentally, many of the best greens are ready to eat in March and April, right when the Lenten fast forbids many animal foods.

One of our discussions with the Psilakis took place on the breezy porch of our hotel, one hot summer evening. Maria, a teacher and effortless polyglot, did most of the talking, her face and gestures expressing as much as her soft voice; she brought along some wild greens and cultivated herbs to share with us, pinching each one, smelling or tasting them herself before handing them over to us. Nikos, although he spoke little English with us, commanded an extensive knowledge of botanical history and nutritional studies, which Maria would translate for us from Greek. We first talked of the *aghria horta* and *votana*—the sour greens and bitter herbs that they brought along for us to sample.

"Almost all of the old people in Crete can recognize every one of the *horta*," Maria explained, "the many species of wild greens. Well, they couldn't have survived otherwise. You must understand that the consumption of wild greens is part of our larger history. Because Cretans were warred upon and subjugated to other rulers for so long, there was often hunger—how do you say it, famine?—in the mountain villages. Whenever they were faced with this hunger, that's when they had to rely exclusively on wild greens. It was because the soldiers would come and take all of their crop harvests and their livestock."

Nikos broke in to her explanation to offer a personal testimony. She listened, sighed, then translated for us from Greek.

"This is why Nikos's grandmother taught her family to survive on wild greens. When the Germans occupied Crete and there was nothing else to cook, she raised a family of seven almost on greens alone, for the Germans had taken all of their other food." In Greek, Nikos sang out the litany of greens associated with each season, but Maria had difficulty translating them all into English. To help Maria, Nikos tried out his English to make his point another way. "Well, the old people of Crete, they knew every place to look for them in each season. I think you can find some kind in some place all around the year."

This reminded me of the remark I had heard a week earlier, when Laurie and I had visited the mountain village of Kastanies on the island of Kéa, closer to the mainland of Greece. We were visiting a farmhouse kept by a woman named Cleopatra, who had recently retired back to Kéa after running a restaurant in a Canadian metropolis for years. Once again, she was living in a landscape filled with familiar wild foods, and with little prompting, her stories about them from childhood spilled out of her. I shared what she said about her mother-in-law: "As we go up from the beach, his mother would yell to us, 'Stop at this spot, I need to pick some plants! Stop at the spot up there too. I want to get a special kind of *horta* that grows just in that place!' "

Nikos nodded his head in agreement. "It is just like that," he said, "only worse on Crete. The Turks say, 'Put a cow and a Cretan in the same pasture, and they will fight. They will compete to see which eats the most *horta*, and usually it will be the Cretan!' "

Maria and I chuckled, though I am sure she has heard the proverb from Nikos a thousand times. Then she grew serious.

"Sadly, only a few of the *aghria horta* are known today among girls the age of my daughter (the one who is in the university studying archaeology), unless their mothers or grandmothers take time to specially teach them by taking them out into the country. My daughter, she knows quite a few now, but that is because we have researched them so long and have a garden full of the ones we are trying to bring into cultivation. But look what the scientists still say about the Cretans—we eat the largest amounts of wild greens of anyone in the world."

Dr. Kafatos confirmed these facts. "We have now recorded 150 species of wild plants still used by Cretans as herbs and greens, and have sent off 65 samples of these *horta* to the University of Vienna for nutritional analysis."

The preliminary analyses indicate that many are excellent sources of antioxidants, folic acid, omega-3 fatty acids, and vitamins. But as Maria reminded us, it was the flavor not the health value of these greens, that attracted the inhabitants of Crete to them, at least as far back as the Minoan civilization.

"We have always relied on them, clear back to the old times. We didn't think of their medicinal properties, we were just attracted to them as soon as they began to come up with the rains in the winter.

Before Easter, during the fasting for Lent, that's when many of the wild greens are at their best.

"When I go out by myself to pick the *horta* near the end of winter," she continued, "I take along with me two bags: one bag is for the bitter greens we put in salads or eat as boiled greens, and the other bag is for greens we mix with meat or cheese in pies, or serve with fish and snails. We never put bitter herbs—the true *aghria horta*—in the pies we call *hortopita*—you know, like how we make *spanokopita* from the *spanaki*. The greens different Cretans favor for their pies can be the stinging nettle, the wild fennel, parsley, wild carrot, or salsify. Then in the summer, we use the cultivated greens fresh for salads and boiled, like the amaranth we call *vlito* and the deadly nightshade we call *stifno*. We have less wild greens and more cultivated ones in the summer, but there are some wild ones from midsummer all the way to autumn, like the purslane."

Our conversation traveled on and on—through delicious marketplaces, lush gardens, and fertile fields, and with every word from Nikos and Maria, I realized just how interconnected all the elements of the Cretan cuisine really are. The mix of plants at the islanders' disposal—we cannot replicate it in too many places around the world. The powerful antioxidant combination of olive oil and greens—we cannot ensure that other ethnic populations will take to these foods culturally as the Cretans do. The dedication to religious fasting, a time-tried way to get clean with God, and clean with our bodies—not many Americans have the spiritual will to fast more than a hundred days a year. And finally, the gene frequencies the Cretans carry—Cretans did not gain their tolerance to imbibing large doses of olive oil, or to resisting malaria through fava beans in the spring as we saw in the previous chapter, overnight.

There is *context* to the way the folks of Spili live and pray, eat and fast, that cannot just be extracted and plopped down in another land to gain the same benefits. We cannot facilely assume that their cuisine will do as much for our genotypes as it does to their genotypes. It is not random; it is embedded in place. And while other Mediterranean dwellers may claim that their diet is purely a matter of taste, there is something deeper and more functional hidden in these gene-diet-culture interactions.

To find out just what that might be, we will have to leave Spili and the surrounding mountains of Crete far behind and travel to the continent that lies on the other side of the Atlantic. There, we will trek across scorching hot regions of the Americas, where a native spice will make us eat our own words the next time we blurt out, "It's merely a matter of taste."

CHAPTER FIVE

# *Discovering Why Some Don't Like It Hot*

## Is It a Matter of Taste?

WE ARE SAILING on to the Americas, following the route that a paradise-seeking explorer named Columbus took on his second voyage to find the pungent spices of what he thought were the Indies. But when he encountered what the inhabitants of this newfound land considered to be their most cherished spice, his physician had difficulty describing it in terms other than its sharp taste. By 1493, his chronicler Peter Martyr called it a "pepper." Whatever the identity of this piquant fruit was—probably a *Capsicum annuum* cayenne variety, the progenitor of most chiles eaten in the United States today—it was certainly not the black, white, or red peppercorn of the genus *Piper* that had been traded across Indonesia, continental Asia, and into Europe and Africa for centuries.

Language often fails us when we attempt to describe the taste of a spice previously unfamiliar to us. But it fails us in a larger sense as well, for we hardly have a vocabulary to explain how we select *any* of the foods we eat. It is clear that most of us are not consciously choosing what we put on our plates by using some data bank that details the composition of a food's macronutrients, its secondary compounds, and how they interact with our genes. Instead, we unconsciously select our meals based on seasonal availability, cost, and, of course, "taste preferences."

But whenever we speak of "taste preferences" or "the foods we choose to eat," we do so as if unbridled free will is the only factor leading us toward the blue-plate special and away from the smorgasbord. Instead, it may be that we are predisposed genetically and culturally to favor the flavor of some foods over others, even though many food aficionados find genetic programming to be a hard concept to swallow. For them, "a matter of taste" literally means a matter of conscious individual discretion.

And yet, taste is something far more slippery. It is a murky realm in which biology, culture, and individual experience come together—complementing one another or sometimes clashing—in mysterious and often painfully amusing ways. Nowhere are the mystery, pain, and pleasure more mixed than in those pungent peppers that Columbus took back from the Americas, the chiles that we will sample on their original home ground, in Mexico and the adjacent U.S. Southwest. There, they are a sensation that literally makes or breaks certain human relationships.

A case in point: I once found out the hard way that the woman I was courting was a supertaster. You need not know much about supertasters for the moment, only that they are hardwired to experience intense burning in their mouths from eating any piquant chile pepper.

And that is why disaster rather than romance emerged when her genetic intolerance of spices encountered my naïveté in assuming that anyone could tolerate as much chile pepper as I could.

It all started with my reading of the Spanish edition of *Como Agua Para Chocolate*, followed by watching the fine film *Like Water for Chocolate* that Laura Esquivel's husband based on her book. Set into the narrative was the meticulous preparation of a complex recipe for *chiles en nogadas*, a dish blending nuts, chiles, and cheese in a manner reputed to produce aphrodisiacal properties. Having just returned from Mexico with many of the requisite ingredients, I thought there would be no better main course for a meal to romantically reunite me with my sweetheart.

After hours of preparation and baking, I was pulling the concoction from the oven to let it cool, when I was called by my partner to leave the kitchen and come outside. There, we glimpsed the last crimson colors of a fiery Arizona sunset and let ourselves cool down on a patio overlooking my garden full of chiles and culinary herbs. As soon as I poured us each a margarita made with homegrown limes and bootleg mescal, I got the first hint that my companion's tastes might not be the same as mine. She took one sip, then pushed her glass away.

"Hmmm, margaritas are something new to me, so if you don't mind, I think I'll pass on this one. The lime juice tastes a little bitter—are you sure you didn't mix grapefruit juice with it? And that bootleg mescal of yours—whoa!—it really scoured my throat!"

"That's OK, let me get you a glass of white wine. You'll need something with the main dish to wash it down."

I went back into the kitchen and brought out two plates covered with poblano chiles stuffed with a special *picadillo* filling and smothered in a creamy almond and walnut sauce, topped with a sprinkle of pomegranate seeds. As my partner lit a candle on the table, I went back in a second time to fetch her glass of wine. We toasted to our reunion, and then I paused before sampling my own plate, set on watching my guest savor her first taste of *chiles en nogadas*.

But she did not take a bite and then move her hand across the table to warmly touch mine. Instead, it was if a fire alarm had sounded. Coughing, she jumped from her chair and ran inside to the sink, where she began gulping down water straight from the tap. When I ran to ask if she had choked on some food, she did not hear me—her head was still immersed in the sink where she was dowsing her lips, her tongue, and her throat with ice-cold water. When she reemerged, she snapped at me.

"I hope that wasn't supposed to be some cruel joke!"

"Wh-whaaat?" I was horrified.

"To make your Mexican food that hot! Did you even sample it yourself before giving it to me?"

"Well yes, I did, in the kitchen, just before serving it. Maybe it hadn't cooled down yet. . . . "

"I'm not talking about the temperature. I'm talking about the chiles! Even my earlobes are burning. Were those the hottest chiles you could find?"

"No, not at all. . . . I mean, I didn't use bell peppers, I just made it with poblanos like the recipe said." I was feeling pretty defensive.

"Look, if I really wanted to burn you, I wouldn't have gone to all this effort to remove the seeds and veins. That makes them less *picante*."

"I think you'd have to get rid of the chiles entirely for me to be able to eat this dish!"

"Well, sure, I could do that," recovering slightly from my frustration. "I suppose I could just give you the sauce and the picadillo stuffing, and chuck the rest."

I went back into the kitchen, and sadly unstuffed the chiles I had so painstakingly filled and drenched with a mix of chopped meats, nuts, fruits, spices, and cheeses. I hastily pulled apart another section of pomegranate and sprinkled the bright red seeds over the cream sauce. When I returned to the patio, my guest had calmed down and was sipping her wine.

"I'm sorry I insinuated that you would burn my mouth on purpose," she said quietly. "I can see that you must have spent a ton of time preparing this meal, but it was just too hot for me. Will you go ahead and eat yours?"

I sampled it, and to her chagrin, I consumed it without any smoke arising from my ears. To my palate, it was surprisingly mild as far as chile rellenos go, savory without pungency overriding the flavor. My companion half-heartedly probed the chile-free sauce with her spoon, but when she tasted it, she again came up frowning.

"Ugh, there must have been a little bit of the pomegranate rind in with the seeds. *Something is really bitter*. Do you have any ice cream I can use to cleanse my palate?"

At that point, I knew that any of the aphrodisiacs that I had hoped would fill the atmosphere around us had already volatilized into thin air.

"Well, my dear," I sighed, "there *is* a little ice cream left in the freezer, but I don't think you want any of it. It's the wild chile and vanilla flavor that Eric's Ice Cream specially prepared for me last season. When they liquefied the little green chiltepines in a blender, two of the workers found they could hardly breathe and ended up in the emergency room." At that point, I threw in the towel. "Maybe we should just go into town and get some Eskimo Pies."

~ Until recently, I believed that several dinners as disastrous as this one had all been caused by my own *loss* of capacity to accurately sense the pungency of chiles. I believed chef Julia Childs, who once intimated that chile lovers literally burn out their taste buds, leaving them unable to sense subtle flavors. But as I have learned more about chiles, it seems as though there is no solid evidence that this taste bud "burn-out" actually occurs.

More likely, my companion was a *supertaster* by her genetic profile, and I am a *nontaster* (with minimal reactions to strong flavors) by mine. These labels are now standard in the study of chemical perception, and I will describe their origins a bit later. In addition, my taste buds might have gradually been desensitized over the course of sampling chiles in doses far greater than most Americans would care to do, during my two decades of fieldwork on the ecology and ethnobotany of peppers. If I am correct, my partner's aversion to chiles and my own capacity to tolerate pungent foods are perfect examples of the fascinating interplay between genes, habitat, culture, and individual experience. Not one of these factors in isolation can account for why some like it hot, and others not; all factors thread their way into a weaving made by the warp of nature and the weft of nurture.

I have been heartened to learn that the reasons behind that dinner's culinary dissonance were far more complex than I could have imagined at the time, as indicated by some recent discoveries about the genetics of taste. I learned of them through correspondence with Dr. Linda Bartoshuk, a professor at Yale University's School of Medicine, who has long shared with me a research interest in a vanilloid compound in chiles, capsaicin. Capsaicin, as you may already know, is somewhat of a chemical paradox, for it can both generate and relieve pain.

As Bartoshuk explained to me, "We have been working on connections between taste and oral pain that have health implications. We have come to suspect that the taste system not only serves as sensory [cues] but also serves to inhibit activities incompatible with eating. Oral pain was the first such activity we studied. Taste input appears to inhibit oral pain in the brain. When taste is damaged, that inhibition is abolished and pain phantoms—sensations in the absence of normal stimulation—are produced in supertasters. The people who suffer from this—it's called burning mouth syndrome—tend to be hypersensitive to chiles."

Although I had read her work on chiles for years, I did not know until we began our recent correspondence that Bartoshuk's lab team had elucidated key features of a polymorphism for the ability to taste chiles as well as other compounds. Her lab had discovered the existence of populations genetically predisposed to so-called *supertasting* just a few years ago, even though the initial discovery of *nontasting* dates back to 1931. That is when chemist A. L. Fox was in his lab trying to synthesize PTC, a bitter compound derived from nitrogen, carbon, and sulfur, and some of its white crystals went airborne. As the crystals volatilized, Fox hardly sensed a thing, while his laboratory colleagues all ran for cover, suddenly perceiving a bitterness on their tongues and in their nostrils and mouths that made them squinch up their faces in disgust.

Within the year, Fox described "taste-blindness" in the pages of *Science* magazine, thus turning his own personal handicap into cutting-edge research. Before another year had passed, Ohio geneticist L. H. Snyder determined that taste-blindness was a heritable trait whose expression varied dramatically within and among populations, suggesting implications for the health status of tasters and nontasters alike; in doing so, he, like Arno Motulsky, helped set the foundation for the field that scientists now refer to as pharmacogenetics, the study of genetic interactions with drugs that I discussed in chapter 3.

Snyder and his followers pioneered a rapid-assessment technique—a bite-size slip of chemically treated paper that one puts on the tongue for a moment—that allowed them to sample members of different ethnic populations for their ability to taste bitter substances. These substances used for sampling included phenylthiocarbamide (PTC) (phenylthiourea, or PTU) and 6-n-propylthiouracil (PROP). Soon, geneticists simply divided the world into the haves and have-nots—tasters and nontasters. Those who puckered up over PTC or PROP also responded to a range of bitter herbs, fruits, vegetables, coffees, and condiments in a consistent manner. The intensity of their responses to mustards, kales, curries, pomegranate husks, peppers, and grapefruits can all be predicted by this simple taste test.

Various field surveys—many of them done by the same Anthony Allison who discovered sickle-cell anemia in Kenya—demonstrated great differences among cultures in their genetic disposition toward taste. About 25 to 30 percent of all Mediterranean residents were taste-blind, while only 7 percent of the Lapps, 3 percent of the west Africans, and 2 percent of the Navajos sampled were nontasters. In Asia, taste-blindness varies from 43 percent in India to 7 percent in Japan. Children identified as tasters expressed their extreme dislike for broccoli and generally avoided other bitter cruciferous vegetables and herbs such as cabbage, kale, kohlrabi, Brussels sprouts, and mustards.

Curiously, nontasters who hardly sense the bitterness of crucifers sometimes eat large quantities of these greens. Nontasters can ingest so many goitrins and isothiocyanates that they interfere with iodine metabolism and generate enlarged thyroids with goiterlike symptoms (PROP is actually a drug used to suppress thyroid function). Obviously, this simple taste test has suggested that heritable taste preferences have had unforeseen influences on our consumption and nutrition, and ultimately, on our health status.

Now, enter Dr. Bartoshuk. Long interested in how taste is affected by genetics and disease, she has, over the years, had her students sample various populations for their responses to bitter or pungent chemicals. In general, students found that three-quarters of those sampled in North America were tasters while another quarter were nontasters—as if there were no middle ground. Then, during one of these surveys in 1991, her student Tracy Karrer noticed that within the group of tasters, individuals varied in the intensity of their responses; some consistently responded to certain chemicals as if they were much more concentrated in dosage than what other tasters perceived.

These "perceptual" differences, as Bartoshuk confirmed in 1994, are rooted in anatomical, physiological, and genetic variation among us. This is because the tongues of supertasters are literally tiled from edge to edge with taste buds nested in dense fungiform papillae, whereas the tongues of nontasters are sparsely polka-dotted with taste buds. Bartoshuk surmised that nontasters have two recessive alleles for the genes influencing taste-bud density and PROP tasting, genes that are now known to be located on chromosome 5p15 and chromosome 7. Supertasters probably express two dominant alleles, whereas normal (medium) tasters likely exhibit a dominant allele on one gene and a recessive on the other.

While the counting of taste buds in the fungiform papillae on the tip of your tongue may initially seem as esoteric as counting how many angels can land on the head of a pin, Bartoshuk can quickly convince you that your status as a nontaster or supertaster *matters*. "Sweets, particularly sugars," she says, "are much sweeter to supertasters, by at least a factor of two. Supertasters feel more pain from lesions on their tongue, and that's medically very important. Supertasters also perceive more pain from oral irritants—chile peppers, black pepper, ethyl alcohol, . . . carbonated water, considerably more irritation from these sources . . . [and] if you're a supertaster, you're a super perceiver of fat in food!"

These perceptual differences predispose nontasters and supertasters to respond with pleasure or pain in varying intensities when exposed to different ethnic cuisines. However, one's previous individual experience, aesthetic biases, and cultural contexts can shift the balance between pleasure and pain. In other words, we can learn (to some degree) to treat the same plate full of *chiles en nogadas* as a piece of culinary artwork capable of providing intense pleasure, or as a sadistic plot intended to inflict pain.

Nevertheless, the health implications of PROP-tasting genes are perhaps even more remarkable than their aesthetic implications. There is a higher proportion of nontasters among families prone to alcoholism, and supertasters are more frequent in families that tend not to find pleasure from the oral burn of strong alcohol. Supertasters tend to dislike bitter grapefruits that are rich in naringin, a chemopreventive compound that reduces the risk of some kinds of cancer. Nontasters are prone to eat more grapefruits, more bitter greens, and more chile peppers, and are therefore offered more protection from some but not all cancers—there is actually some suggestion that chile eating is correlated with higher risks of stomach cancer.

More startling still is the finding from Bartoshuk's lab that elderly women who are supertasters tend to weigh less, carry lower percentages of body fat, and have lower triglyceride levels as well as higher levels of the "good" cholesterol, HDL. So while women who cannot stand bitter greens and chiles may be exposed to the risk of some cancers, they are clearly not as much at risk for heart disease or alcoholism.

Beverly Tepper, a food scientist at Cook College at Rutgers University, has neatly summed up the implications of those genes on chromosomes 5 and 7: supertasters, medium tasters, and nontasters truly live in different "taste worlds." This predisposes each group to different health hazards and benefits, to different experiences of pleasure and pain, and to different reactions to chile peppers. Realizing that these groups are probably not equally distributed across the face of the Earth, Tepper and her colleagues have laid out an intriguing hypothesis: "Most chile lovers would be expected to be nontasters. [And so], if the liking of chiles is closely linked with PROP-taster status, then areas of the world where chiles are widely consumed would have a high frequency of non-tasters in the population" (Tepper 1998).

The hypothesis remains viable, although Bartoshuk was quick to remind me that the sufficient data to fully prove it are not yet in: "Since the discovery that there is an association between the genetic ability to taste and the perception of the oral burn of chiles is a rather recent one, I don't think there have been any studies of geographic distribution based on perceptions of oral burn . . . the discovery of supertasters is too recent for that to have influenced field studies yet."

Of course, that may be changing as we ponder Dr. Bartoshuk's very words.

~ It is time that we visit a place not far from where Hernando Cortés made his own landing when he came to conquer Mexico, and where chile-loving nontasters are surely found in abundance. The place is Xalapa, Vera Cruz, the area of Mexico that gave a gift of inestimable value to the rest of the world: the jalapeño chile pepper. In Xalapa, another Rutgers scientist, Beverly Whipple, led local women into a rather saucy study of the effects of chile eating on the experience of sexual pleasure and pain (Whipple et al. 1989).

For several years, Whipple and her colleagues had been studying the analgesic (pain-relieving) effects of genital self-stimulation, but they realized that certain factors could override the sense of pleasure that some women might otherwise obtain from orgasm. The researchers guessed that chronic chile eating might "swamp" the endorphin system in the body so that chile eaters might be less tolerant of pain. But to confirm that hypothesis in 1989, Whipple undertook one of the most bizarre studies I have ever read, one that went to the heart of chile-eating country in Xalapa.

Imagine trying to recruit twenty-five Mexican women between the ages 22 and 50 so that you could study the effects of their vaginal self-stimulation on pain relief. You must first find women who are even willing to participate in an activity that fundamentally challenges their sense of privacy and sexual intimacy. Then you must take them through a rather tedious battery of questions about their diets and their mental health. Do you love chiles or hate them? Do you eat them regularly? Which kinds then, only little-bitty mild ones or really big hot ones as well? By the way, have you ever needed a psychiatrist?

You then divide the women into three subgroups, the first self-identified as chronic, high chile consumers; the second eating a medium chile diet; and the third being chile averse, with little hot food in their diet. Although we will never know for sure, since supertasters had not yet been described, perhaps the first group had more non-tasters and the third group had more supertasters.

At this point in the study, you let each individual from all three of the groups relax in a reclining chair, and you encourage her to stimulate herself with a furry mitt on her right hand, while holding the other hand out where the sharp pinlike point of an "Ugo Basile analgesia meter" can be slowly lowered down upon it—a standard means of measuring pain tolerance. The deeper the pin pressures into the subject's flesh before the pain is too great and the individual yells "*basta*!," the higher the pain-tolerance rating.

And so, Whipple and her colleagues in Mexico confirmed their hypothesis—that the women who chronically consume the most chiles accumulate so much capsaicin in their bloodstreams that their responses to pain are no longer buffered by the analgesia produced during orgasm. Their endorphin engines have already been flooded—that is, supersaturated—and these women hardly respond to their ignition switches being triggered. In contrast, the chile-averse women—most of them who are probably supertasters—experienced the full analgesic effects of self-stimulation, with pleasure masking pain.

Forget for the moment that the study had anything to do with furry mitts, self-stimulation, and orgasms. Instead, take the upshot of this study to an emergency room in any sizeable North American city where a great admixture of ethnicities lives in proximity to one another. As any emergency-room nurse can tell you, individuals of different ethnic origins react in decidedly different ways to the pain that accompanies accidents of various magnitudes. Although there is always the risk of inappropriate ethnic profiling by taking quick snapshots at an emergency-room scene, it is possible that such snapshots sometimes catch people engaged in their most visceral responses to the world. And although these responses might be in part culturally programmed, other elements may be due to gene-diet interactions.

Now, let's go north, to the multicultural Southwest of the United States, a place still within walking distance of wild chiles, but one now inhabited by many cultures other than those of pre-Columbian origin. Imagine sitting with me in an emergency room in Tucson, where I waited to be examined for a hyperallergic reaction to eating watermelon that had nearly swollen my windpipe shut. I kept panting, wondering when on Earth a doctor would see me.

I had some time on my hands, so I stared at the other patients. Sitting stoically beside me was a big Navajo Indian cowboy, who was bleeding into a towel pressed over his abdomen. When I asked him what had happened, he was almost inaudible. He muttered that while sitting at a bar having a beer after work, someone came in yelling, pointed a shotgun at the guy on the barstool next to him, and when the gun went off in a struggle, this Navajo man had been sprayed with some of its pellets. He sat there, holding his stomach and sipping water until the nurse called him.

I could hardly hear his story though, for next to him were a daughter and a mother from a Mexican family I had met at several times at a local tortilla and tamale factory nearby the hospital. While preparing chicken for some *mole pipian*, the daughter had accidentally cut off the tip of her finger and was constantly crying with pain. Her mother was all over her, holding her, drying her tears with a moist rag, and praying to every saint she could think of while they waited to get in for treatment. The daughter would mostly whimper, but when the throbbing of her finger would overwhelm her, she would let loose with a loud *grito* as if it were Mexican Independence Day.

The mother looked somewhat annoyed when an Italian man in his midthirties came in with his entire family and made her move a purse off the seat next to her, to clear the way for his kin. This *Italiano* was howling with pain and continually asked the nurse when he could see a doctor. The Mexican mother told me that they knew him from their church and that he too was a cook, at the parlor that made the spiciest pizza in town. Apparently he had not been working that day, but was doing some carpentry at home when a large splinter of wood jabbed in deep under his thumbnail. As he screamed curses in Italian, his kids all cried, convinced that he would die on the spot. He made so much racket that a doctor came running, tugged at the man's arm, and they headed off to a private room. I could hear the man's curses echoing all the way down the hall.

Obviously, there is no way for me to know who among this group are tasters or nontasters, or even who if anyone had recently eaten chile in any quantity. But whenever I am catapulted into scenes such as this, they do not seem like blank slates any more; I wonder who among us is genetically predisposed to dramatically expressing their pleasure or their pain. Cultural, culinary, and genetic factors may strangely come into play, as one person or another is inclined to howl, to whimper, or to stoically sit there waiting one's turn.

Pain and pleasure, peppers sizzling, and the sensation of burning. They are all weirdly related to one another in our hearts, minds, and stomachs. In the heat of passion, we suffer from broken hearts and from heartburns, but some of us still love saucy, red-hot mamas and pit-roasted barbecues that are hotter than hell. We get the hots for someone, but then that person turns around and burns us. Mexican men even nickname their penises for different kinds of chiles, but their women complain when "the fire went out" of their lovemaking. Sometimes, we have trouble discerning whether we have been scorched or whether the relationship is still sizzling—it's all too close to tell.

~ It may be time to ask, why do we feel that chiles are burning us, anyway? After all, they are not literally "on fire." This is a question that has two answers. To understand the *proximate* answer of why we, along with all other mammals, sense a burning sensation in our mouths during the ingestion of chiles, we must first search for the *ultimate* evolutionary and ecological answer to a deeper question: how does a chile plant benefit by making mammals experience a burning sensation when the mammals sample their fruits?

I had the good fortune to be part of the field research team that recently offered the answer to that problem in the pages of *Nature*, the British science journal that loves hot topics. And most of our problem solving was *not* done in a laboratory, but in a rugged canyon near the Arizona-Mexico border, the northernmost locale where wild chiles have interacted with the native fauna for thousands of years. There we could see how piquant peppers evolved their mechanisms of attraction and defense, encouraging some animals to eat them and disperse their seeds, while repelling those that might sample them and destroy their seeds, thereby limiting the chiles' reproduction and survival.

It turns out that nearly all of the wild chile-pepper plants that occur in this canyon are rather tender, gangly plants that must take shelter beneath larger, shady trees in order to protect themselves from the occasional freezes and soaring summer highs of the desert climate, as well as from large herbivores browsing and trampling as they shoulder their way up through the canyon's brush. Curiously, we found that four-fifths of the chiles that survive these perils reside under just one kind of protective cover, the hackberry tree, for it provides the kind of dense, spiny shelter that buffers pepper plants from various stresses.

However, hackberries do not make up a very large percentage of the canyon's vegetative cover; but chile seeds were somehow getting dispersed to these "nurse plants," where they survive with greater frequency than they do under other, more abundant trees in the canyon. And so, a brilliant friend and former student of mine, Josh Tewksbury, decided to set up some video cameras on and under hackberry nurse plants to catch on film just who was moving chiles seeds around.

The videos—corroborated by our own observations through binoculars from the canyon rim—identified thrashers, mockingbirds, cardinals, pyrrhuloxias, and finches as being among the primary chile harvesters and dispersers at these sites. In the late summer and early fall, some of these birds crave carotene, a vitamin found in high concentrations in chiles, for it increases the intensity of the birds' plumage color as they grow out new feathers after their summer molt and before their late-fall migration. Wild pea-sized chiles provide such a carotene rush. But what amazed us the most was that these birds spent an inordinate portion of their time picking chiles, then moving high up in the canopies of hackberries to roost where the sun reaches them over the canyon rim. Their behavior provides a near-perfect means for chiles to be dispersed beneath the hackberry trees, where the seeds shower down to fertile ground for germination. The seeds "shower down" in two ways: first, as the messy eaters above fumble them while eating, and second, a few minutes later, as the birds defecate them out. The ingested and defecated seeds largely remain intact, undamaged, and prone to germination after dispersal.

Although we had confirmed that certain birds served as effective dispersal agents for chiles, transferring seeds to microhabitats or "safe sites" where they could grow best, we could not yet rule out that mammals could do the same trick. But when we set out live traps to survey which small mammals resided in the canyon, few of the trappings clustered around the chiles and hackberries. So Josh and I set out paper plates on the canyon floor, mixing hackberries and wild chiles on them in equal numbers and counting the number of fruits removed the next morning.

The hackberries were sampled and taken away by small mammals, but the chiles—nearly identical in size and color—were hardly touched. After a single fleeting taste, it appeared that the mammals avoided the chiles altogether. In addition, our lab experiments confirmed that chile seeds force-fed to mammals are often destroyed by their teeth and guts and lose their germinability. Mammals, we surmised, were poor at dispersing chile seeds, had no strong predilection for hanging around hackberry trees, and would probably damage any chile seeds even if they somehow ingested them.

Our lab experiments with the wild birds and mammals native to the canyon confirmed what experiments on domesticated animals had predicted: that the birds barely sensed the pungency of chiles, whereas mammals exposed to them showed immediate aversive behavior, and if force-fed them, would not only lose weight, but would sooner or later experience deteriorating health. And so, I came up with a little poem to offer as an answer to why chiles are hot: "So that birds will disperse them to nurses, while other critters will not."

Conceding that my poem was cute but probably not worthy of publication in a science journal, Josh ingeniously coined a technical term to relate this newly described phenomenon to scientists rather than to those who read poems. He called it *directed deterrence*. Chile plants' chemical arsenal of capsaicin effectively works to deter mammals from even trying to disperse chile seeds, since it is unlikely that the seeds would reach safe sites for germination and establishment through this means. At the same time, birds are rewarded for their dispersal of chiles with a needed dose of carotene and other nutrients; they are physiologically undeterred by the consumption of capsaicin.

Within just two years of publishing our explanation of why chiles are hot, Josh and I were delighted to learn that two other scientists had uncovered the molecular basis for directed deterrence in chiles. Regardless of whether or not a mammal is a supertaster, medium taster, or nontaster, their bodies utilize a particular pain-sensing pathway known as VR1 that "reads" the presence of capsaicin in red peppers just as it reads an increase in temperature. It also evokes a tingling and burning pain sensation when exposed to other vanilloids similar to capsaicin that occur in black pepper and in ginger. These irritants, just like the presence of fire itself, activate a dramatic response in our sensory nerve endings and in those of all other mammals tested to date.

Curiously, University of California scientists Sven-Eric Jordt and David Julius found that while something akin to this same pain pathway is present in birds, it fails to be activated by all but the greatest megadoses of chiles; instead of being stimulated by oral irritants, it is usually stimulated only by heat. Because of some small but significant molecular differences in their pain pathway, birds are indifferent to the pain-producing effects of capsaicin and are therefore not at all discouraged from eating and dispersing chile seeds.

Jordt and Julius learned that neither birds, reptiles, nor amphibians have much capacity to chemically sense capsaicin as a source of "heat." Only mammals have the innate "sense" to avoid chiles, and this sensitivity appears to have developed rather recently in their (and our) evolution. The question, then, becomes even more perplexing: why do humans override the inborn "common sense" that we share with all other mammals—so that we go ahead and eat chiles anyway—ignoring hundreds of thousands of years of genetically programmed signals telling us to avoid such inflammatory oral irritants?

To find a satisfactory answer to that question, I left the Arizona canyon full of wild chiles behind and visited one of the foremost thinkers on the biological and cultural roots of food choices at his office in New York City. During his three decades of investigating cultural food preferences, Paul Rozin has often used chile peppers as his case in point. He has suggested that humans first used chiles as a vermifuge and topical medicine before trying them internally; when we were sure that chiles could not poison us, we began to crush tiny wild chiles to "salt" our food with them. Later, we domesticated the larger chiles, which we now use, not only as condiments, but as vegetables in and of themselves. As Rozin aptly summarized that trajectory, "In almost every culture, at least one innately unpalatable substance becomes an important food or drink."

And yet, as I reminded him, chiles have recently become the most widely used spice and condiment in the world, poured on nearly as many plates of food as salt is sprinkled. How do we sort out which of the many benefits offered by chile peppers to its aficionados was the one that overrode an initial aversion to chiles?

"That's what the debate has been about, " he conceded. "The adoption of chile peppers as a food in pre-Columbian American cultures, and its rather rapid adoption by other cultures in the Old World tropics following the Age of Exploration—remain real puzzles."

Rozin has, over the years, entertained several hypotheses for why chiles have trumped our genetically predisposed capacity to feel burned by chiles and to therefore avoid them. For starters, chiles offer novelty in diets where the staple foods are often bland and monotonous. Or perhaps chiles provide a mechanism to maintain emotional homeostasis by restimulating our attention when we become distracted. Of course, chiles have chemicals in them that delay food spoilage, or at least mask the smell and taste of spoiled foods. Furthermore, chiles make the dwellers of scorching climates sweat in a way that cools them off, a sort of poor man's air-conditioning in a fiery red pod. And finally, chiles supply essential micronutrients and protective antioxidants.

After years of puzzling over these riddles, Rozin has seen a flaw common to nearly all of these hypotheses: "The preference for chile peppers is not motivated primarily by a desire for the consequences of eating it; that is, people do not eat it in spite of its 'bad' taste, as they do a vitamin supplement. Rather, they show a truly affective shift. . . . [T]hey come to like the same 'burn' they used to dislike." In other words, though we may be attracted to eat chiles despite the burn, what attracts us may not be linked to the ultimate benefits chiles provide.

Regardless of why people consciously choose to eat pungent peppers, some biologists argue that there is one undeniable benefit that chiles offer, a benefit that increases the fitness of those who regularly consume them. Chiles—as sauces, spicy powders, or whole pods—reduce the voracity of microbes hiding within the food we eat and limit their capacity to poison us.

Cornell University biologists Paul Sherman and Jennifer Billing further contend that the antimicrobial hypothesis works not only for chile peppers but for other meat-seasoning spices as well. In short, spices fight the bacteria and fungi that spoil meat-based food to the point of making us sick or killing us. Sherman and Billing contend that this is particularly true in desert and tropical climes where cooked meats rapidly spoil if left unattended or unseasoned. Like some other spices, chiles cleanse meat of parasites and pathogens before it is cooked and eaten, and chiles contain four kinds of antioxidants capable of repelling microbes even after a dish is prepared: ascorbic acid, capsaicinoids, flavinoids, and tocopherols.

To test their favored hypothesis, Sherman and Billing analyzed cookbooks of traditional ethnic cuisines from northern and southern boreal zones, clear to the equator, assuming that areas closer to the equator would suffer higher ambient temperatures and more-rapid meat spoilage. For each authentic place-based cuisine, they recorded the percentage of meat recipes that included chile peppers and other spices, as well as the number of vegetable recipes containing spices. They predicted that if antimicrobial defenses against meat spoilage were the driving forces behind eating spices, the same ethnic cuisine would contain more spicy meat dishes than vegetable dishes. In addition, the percentage of meat dishes containing spices would increase the nearer the culture lived to the equator.

Their hypothesis held! Analyzing some 4,500 meat-based recipes and 2,129 vegetable-only recipes in 107 traditional cookbooks from thirty-six countries, Sherman and Billing confirmed that thirty-eight spices were used less frequently in vegetable recipes than in meat recipes. They also confirmed that the intensity of use of chiles and other spices is higher in hot climates where meats spoil relatively quickly. Furthermore, chiles and the few other spices that inhibit microbes the most are favored in hot climates close to the equator. Sherman and Billing confirmed from lab studies that the amount of spices used in meat dishes in both the dry and wet tropics are sufficient to kill the particular bacteria and fungi that cause meat to spoil and meat eaters to get sick.

To highlight the advances that this novel microbiological interpretation of ethnic recipes led to, Sherman crowned their work with the title "Darwinian gastronomy." As their paper was published, the editor in chief of *BioScience*, Rebecca Chasan, praised it as the dawning of a whole new field providing a set of ingenious tools and evolutionary principles that biologists could use to interpret human diets just as they do the diets of other organisms.

Chasan wrote, "As any world traveler or adventurous restaurant-goer knows, some cultures make ample use of spices, whereas others use them only sparingly. Indians use garlic, onions, chiles, and pepper in abundance, whereas Norwegians rarely serve highly spiced food. What accounts for such differences in spice use?" (Chasan 1999).

Chasan's enthusiastic answer to this question implied that a single ecological factor—the antimicrobial activity of spices—was the only one needed to explain the evolution of these gastronomic patterns, simply because reducing food spoilage has such survival value. If chiles reduce the frequency of deaths due to food-borne illness, evolutionarily speaking, the pods were worth their weight in gold prior to the availability of refrigeration. But what the larger picture shows is that one gene for tasting and other genes for an ancient pain-sensing pathway interact with environment, culture, and behavior to shape the degree to which each of us takes pleasure or pain in the eating of chiles. Even Sherman, the man who coined the term Darwinian gastronomy, conceded that relatively rapid "non-Darwinian" processes might be involved in the coevolution of chiles, microbes, and human cultures.

"Over time," wrote Sherman and his student Billing, "recipes should 'evolve' as new bacteria and new fungi appear or as indigenous species develop resistance to phytochemicals, requiring the addition of more species of new spices to combat them effectively. . . . Thus, cookbooks from different eras are more than just curiosities. Essentially, they represent written records of our co-evolutionary races against food-borne diseases" (Sherman and Billing 1999).

In that mouthful of three sentences, Sherman and Billing are acknowledging the possibility that certain evolutionary processes may be changing gene frequencies of fungi, bacteria, spice plants, and humans as they struggle with one another in an evolutionary arms race, all of which raises further questions. For example, since chile peppers did not reach India until some time after 1492 AD, the meat-infecting microbes in that region have only been reacting in the arms race with chiles' antioxidant compounds for some five centuries, while such microbes have had millennia to respond to the chemicals in Old World spices. Paul Sherman's hypothesis would suggest that chiles would have joined other spices in Indian recipes for meat dishes more immediately than they did for vegetarian dishes. In other words, the addition of a new microbe-deterring spice to a meat-based cuisine would have had immediate adaptive value to its consumers, since the microbes may have already developed some resistance to the chemicals in spices that have been part of that cuisine for a longer period of time.

In a similar vein, the Columbian introduction of livestock to Central America undoubtedly increased the amount of meat that Neotropical microbes could infect in Mesoamerican households. Did this lead to the selection of more pungent chiles to sprinkle onto beef jerky as it dried in the hot tropical sun? Did nontasters or supertasters have greater chances of survival as the dynamics shifted toward more meat eating, more pungent peppers, and more virulent microbes?

While these historic phenomena have yet to be studied in any depth, we can be sure that some principles will stand the test of time: First, why some cultures like it hot while others do not is due in part to the nonrandom distribution of meat-spoiling microbes and antimicrobial spices around the face of this Earth. Second, because the distributions of particular bacteria and fungi, chile peppers, and ethnic peoples continue to change, neither the frequency of chile use in "traditional" ethnic recipes nor the frequency of nontasters in an ethnic population are likely to remain static.

As we shall see in the following chapters, there have been unprecedented changes in the geographic distributions and population sizes of different ethnic groups over the last five centuries; and as people have left their ancestral homelands for new habitats, they have brought some of their traditional plants and animals with them to create what historian Alfred Crosby calls "ecological imperialism" in the landscape and in the lunch room. Migrating peoples have also been exposed to new plant and animal foods, new parasites, new diseases, and new stresses, many of which their genes were not necessarily preadapted to deal with. And so, the some 2 billion people on this planet who have been either willing immigrants or unwilling political and economic refugees are now responding—genetically, ecologically, and culturally—to unforeseen evolutionary pressures. The issue facing us today is whether sufficient adaptations to these "new" pressures are emerging rapidly enough through natural and cultural selection and through technological fixes so that humankind in all its diverse forms can maintain resilience in the face of such accelerated change.

CHAPTER SIX

# *Dealing with Migration Headaches*

## Should We Change Places, Diets, or Genes?

BECAUSE of where your ancestors lived and what they ate, your genes interact with the foods that both you and they have eaten—as well as drinks and drugs you've ingested—in profound ways. Then comes the puzzling truth: however profound, those ways remain hidden from your sight most of the time. In fact, these interactions often become apparent to you only because your current place of residence and diet may have become out of sync with those of your ancestors.

From places we have visited thus far, it may already be evident that our genes have set us on certain time-tried paths, but they are paths that branch and meander. As an entire people migrates away from their former homeland, the displacement changes who they are, both genetically and in terms of their cultural identity; and earlier stretches of their collective trail may become obscured. This implies that factors other than genetic ones can alter or mask heritable effects to the extent that we cannot necessarily discern if our genes are leading us back home to ancestral lands or far away from them. Whenever an ingredient of our diet or medicine cabinet triggers a basic genetic response from our bodies, our current habitat of residence, cultural practices, and individual behavior fill in the details. These details may flesh out a picture much different from that produced when our great-grandparents' bodies interacted with *their* homeland and hearth.

Try to imagine all the toxins in and around our kitchen and garden to which we are potentially exposed, for most of our plant and animal neighbors rely on *chemical* defenses more than they do spines, stickers, teeth, and claws. Wherever we are, these natural poisons lie in wait. They are in the juices we drink, the peanut butter we smear on a piece of bread, and in the mold growing out of the very underside of the bread. They are in the herbs imbedded in a piece of cheese, and in the cheese itself. They are in the tapioca pudding, and in the dessert wine we leisurely sip after dinner. Fortunately, most cultures have found means of preserving and preparing foods that may detoxify or at least dilute these poisons.

Furthermore, our ancestors may have habitually consumed potentially toxic foods in conjunction with herbs, brine, fruit juices, vinegars, or other natural additives that nullified or counterbalanced the most hazardous chemicals in them, chemicals that would have otherwise wreaked havoc with their health. Some of the most curious means to keep plant toxins in check are widely practiced even without the understanding of *why* they work—people simply know that such methods do indeed work. Think of a peasant farmer in northern Italy or Morocco soaking olives in a thick brine or drenching them with lye; these farmers do not know what chemical reactions ensue that deactivate the bitter glucosides in freshly picked but uncured olives; all they know is that either of these curing processes works to make the olives edible, or in some cases, delectable. The same can be said for those African and Latin American women who process bitter maniocs by adding their own saliva to the root pulp, leaving the mixture to sit in vats or bowls until it has turned sweet. The detoxification of otherwise nutritious plant foods did not begin in some antiseptic laboratory run by PhDs in chemistry; it began at the hearths of farmers and foragers who learned by trial and error to transform the value of these plants using other potent substances found well within their reach: salt water, clay, fermented fruit juice, and saliva.

The chemical processes of detoxification may be complex, but nutritional anthropologist Tim Johns has found them to be practiced across countless ethnic traditions. The custom of eating clay with potentially poisonous wild potatoes, for instance, is common wherever native potatoes grow in the wild or on field edges, from the Quechua in the Andean highlands of Peru, northward to the Diné or Navajo of the American Southwest.

It is to the land of the Navajo that our story leads us next, where we will see what happens when an ethnic population migrates far from their ancestral homelands and intermarries with another people already residing in that land. Although the Navajo and their Athapaskan kin have lived in North America for at least fifteen centuries, the Navajo frequently moved from one habitat to another until they finally came to reside around the Painted Desert of the Four Corners states. The Navajo have certainly adapted in many ways to their current home and its foods, while retaining many of the genes that they have long carried. This is the story of how some rather simple nutritional interventions have reduced health risks that the Navajo might otherwise have experienced in their newfound lands. This story, in many ways, foreshadows the challenges faced by some 18 million people on this Earth today who have had to emigrate from their original homelands because of war, terrorism, disease, or economic misfortune. But it is also a story that hints at why new gene therapies might not necessarily be the best answers to the particular challenges that these peoples face in their more recently adopted homes.

~ Navajo medicine man Mike Mitchell appeared at my office door one day, wearing a big, cream-colored, straight-rimmed cowboy hat, high boots, and faded blue jeans, with a stunning necklace of turquoise nuggets slung around his shoulders. I welcomed him in, for a mutual friend had suggested to each of us that we should talk. To say that there was a twinkle in his eye is understatement; it seemed that just about everything we discussed that day amused Mitchell. Herbalist, oral historian, educator, singer, and sheepherder, Mitchell had seen the world from various vantage points over his seven decades of making a living in northern Arizona, and much of what he now saw made him smile. On that particular day, he was chuckling over the notion that the college professor whose office he was in was trying to raise Navajo-Churro sheep! How in heaven, I am sure he wondered, could a person be a decent sheepherder if he was stuck in an office for most of each day?

I told him that because of the lack of forage from a recent drought, we had all of our sheep corralled and were feeding them some rather expensive timothy hay. And because five of the ewes were bred last fall and were close to lambing, we were keeping a close eye on them in the corral.

He explained that when there were droughts such as this one, his people herded their sheep up into higher country where the animals could browse on a more steady supply of big sagebrush, what the Navajo call *ts'ah*. Such a diet in the months prior to being harvested, he added, gave the mutton and lamb the fragrance of sage.

"We don't have to add anything to the meat when we roast it. It's got the herbs right inside it," he said, licking his lips.

"There's a similar thing with the lamb where my people come from on the border between Lebanon and Syria. You know, the other Holy Lands," I replied, his story stirring a memory from my last visit to the ancestral home of the Nabhan clan. "The high meadows in the mountains there are loaded with wild thyme, so people graze their sheep up there. My people like sage, too, so they make slits in the meat before they roast a leg of lamb, and they stick sprigs of the sage into it like that. Do your people ever use sage as an herb to flavor anything else?"

Mitchell chuckled again, the deep sun-worn wrinkles around his eyes arching upwards. "*Ts'ah*? We take it with many different things.

It's medicine, and it's food, you name it. Sometimes we just pick off a twig while we're riding along and chew it. We use its pollen, you know, plus its leaf, its root. You like that plant? Let me show you something I wrote up for our children, so they can learn their own traditions in the school."

He pulled out of his satchel a thin but beautifully illustrated booklet on Navajo uses of plants that he had helped prepare with teachers in Chinle. He turned to the pages regarding the plant he calls *ts'ah*, written in both Navajo and English: "The strong odor of the sage plant is unmistakable and considered by many to be rather pleasant. It produces light green pollen. This plant is used extensively for flavoring food. When forage is scarce during winter months, livestock browse these plants" (Mitchell et al. 1998).

I suddenly realized I should have brushed up on the knowledge of my Navajo neighbors as soon as I brought sheep to my homestead the year before. I read on: "This plant became a Life Way herb, when an ill person happened to be near sagebrushes. Traveling alone, a sick man came upon an aromatic shrub, examined the plant, and noted that the sage aroma made him feel better. The leaves tasted pleasant, so he began chewing on them. As he sat in the shade of a sage shrub, he felt a complete healing. A Life Way medicine was discovered."

Mitchell interrupted my reading to explain how Navajo medicine men classify the many plants they encounter into four main groups: foods, many of which also have medicinal properties; medicines; plants of no currently known benefit; and poisons. I was initially puzzled by these four groupings, since I assumed that many of the medicines, at high dosages, might be poisonous as well. I stumbled for a moment, unsure how to phrase my question.

"Can someone get even sicker . . . or even get poisoned if they don't use the medicines in the right amounts?"

Again, Mitchell chuckled quietly. "That's the trouble. Sometimes the people just go and get their medicines from the flea market without talking to any medicine man. They don't know what to do with it, so they end up in some hospital. Then the hospital tries to find me to see if I know why. Why these people get even more sick after they take the plant medicine. Those people tried to make something with the medicines, without knowing how, they put in too much."

~ To a Navajo medicine man, healing is nothing if not deeply contextual. How a medicine is prepared, how much of it is administered and when, and who it is given to, are every bit as significant as the medicine itself. And that is what is interesting about the deep bond that the Navajo have with sagebrushes, which are in the genus of shrubs and herbs *Artemisia*, a group of plants whose members have been found to be ritually placed in the oldest known human burials. Sages are pharmaceutical storehouses for a class of lactones known as coumarins. Lactones are cyclic esters of hydroxyl acids, formed by removing water from the molecules of these acids. And coumarins are, biologically speaking, as physiologically active as any ingested substance can be.

At different dosages and in different contexts, coumarins function as *chemopreventives*, plant chemicals that can prevent diseases or control other stresses to the human body. Chemopreventives such as coumarins can kill nematodes and worms, repel insects, discourage the growth of molds, and inhibit the seed germination of potential plant competitors. Of course, they can also function as flavorings, and as fixatives for the fragrances of perfumes. Somehow, coumarins can lower the probability of addiction to nicotine, thereby reducing the risk of lung cancer, but at the same time, they may directly stimulate the growth of certain other cancers. In fact, some coumarins hold the dubious distinction of being the most potent natural carcinogens with which humankind must deal. Then again, certain coumarins historically served as narcotics and as sedatives. They are still used to stimulate or to depress the central nervous system, to eliminate intestinal parasites, and to speed up or to stop blood clotting. Depending on who and where you are, coumarins can be either a blessing or a curse.

For a group of compounds as biologically active as coumarins can be in the human body, it is intriguing just how many foods and medicines contain them. If these foods and medicines were not anciently used but new to humankind, and the Food and Drug Administration had to determine whether they were safe enough to put on the market, I doubt whether the agency would approve them. The reason is that they can trigger so many deleterious as well as beneficial effects in human metabolisms.

Nevertheless, medical anthropologist Richard Raichelson has estimated that nearly half of all the plants traditionally used for food, medicine, and personal care by all cultures in the American Southwest may potentially contain coumarins. In a more detailed analysis of plants used by the Navajo, Raichelson confirmed that as much as a quarter of all their native foods and medicines contained coumarins. Wild potatoes, sumac berries, juniper berries, and purslane leaves are among the commonly utilized, coumarin-laced foods and medicines of Navajo country. To be sure, coumarin-rich plants are not restricted to the deserts and shrubby steppes of the American Southwest. Among the coumarin-containing plant foods found in your grocery store are chickpeas, parsnips, coriander, cherries, plums, oranges, and figs.

If coumarins are so widespread throughout the plant kingdom, you may already be wondering how they could pose any special risk or benefit to Navajos such as those regularly treated by Mitchell in his work as medicine man. That is where the story gets interesting; in fact, exceedingly interesting, considering how intensively the Navajos use coumarin-rich plants. Navajo Indians are one of the few ethnic populations known to carry two of the variants of serum albumin A, both of which make carriers acutely responsive to ingested coumarins. This polymorphism potentially affects the health of the Navajo in numerous ways.

If you go back to your notes from Blood Chemistry 101, you will recall that albumin is a soluble protein that comprises about half of all the protein in your blood serum. It is a *carrier protein* that plays a key role in lugging around fatty-acid molecules, steroids, and thyroid hormones in your bloodstream. Drugs get bound up and deactivated or further circulated, depending on which variant of serum albumin you happen to have. If you carry one copy of this allele from both of your parents, you are homozygous for the "typical" albumin A. That means that more of a drug such as warfarin (an anticoagulant) would be fully bound up and deactivated in your body, making blood clotting normal.

But if you happen to be heterozygous for one or the other of two allelic variants of albumin A, *Mexico* or *Naskapi*, about 27 percent more of the warfarin drug would stay active in your bloodstream, so your clotting would become impaired and hemorrhaging more difficult to control. Such a result means that you have a mutation of the albumin gene on chromosome 4 that generates various anomalous proteins, all found in smaller quantities than those in typical serum albumin. Just as warfarin would have more potent effects for you, so would the naturally occurring coumarins in plants such as sagebrush. Medical anthropologist Nina Etkin puts it this way:

> The significance of such variability is illustrated by investigations of coumarin-containing plants used by Native American populations who have relatively high frequencies of serum albumins "Naskapi" and "Mexico." . . . Insofar as these bind significantly less synthetic coumarin (warfarin) compared to ("normal") albumin A, individuals with albumin variants have more active coumarin in circulation. This has important implications for the use of warfarin in Western biomedicine, since dosage may have to be varied in accordance with the drug-binding capacity of different albumins. Further, these individuals may vary as well in their interactions with other (natural) plant coumarins with which they come into contact . . . i.e., the therapeutic, toxic, and other effects of various coumarins may be exaggerated in individuals who have these albumin variants (Etkin 1986).

Many Navajos carry the Naskapi variant that is also carried by their Athapaskan kin in Canada and Alaska. This Naskapi allele is also found among the Eti Turks of southeastern Turkey, with whom the Navajo may have shared ancestors before the Athapaskans migrated out of Asia thousands of years ago. Both groups—the New World Athapaskans and the Eti Turks of the Old World—may have had roots in the cold desert steppes of central Asia, before moving eastward and westward, respectively. We still do not know for sure how closely related they are, but when these peoples meet, they are immediately struck by how much their languages and their physical appearances have in common.

~ There lies the irony: they left one coumarin-rich landscape for another. Somehow displaced from central Asia, where their languages branched off from the Sino-Tibetan tongues still spoken there today, the ancestors of the Navajo first migrated eastward into Alaska and northern Canada, then southward into the intermountain region of North America. A thousand years ago, they ended up in a semiarid land as rich in coumarin-laden plants as that from which they emigrated. Many of these plants triggered, then and now, unusually strong responses whenever ingested in any quantity—the coumarins are potent medicines, but are also potential poisons and carcinogens.

The Athapaskan emigration out of central Asia into the American Southwest did not merely expose the migrants to more coumarins, it also allowed them to pick up another variant of the polymorphic albumin A gene. Some Navajos now carry the Mexico variant that is also commonly found among speakers of Uto-Aztecan languages in the American Southwest and western Mexico. After the Pueblo Revolt of 1682, the Navajo frequently took in and formed bonds with members of Ute, Pima, and Hopi tribes of the Uto-Aztecan language family, who had already been living in habitats covered by coumarin-rich vegetation. The Mexico variant likely entered the Navajo population from intermarriage with these tribes.

And so, in their newly adopted home in the American Southwest, Navajos encountered both more coumarins *and* a second albumin variant that conferred additional susceptibility to potent coumarins. Genetically speaking, this is a double whammy, especially for anyone living in a place loaded with plants that can trigger strong reactions. And yet, medicine men continue to use big sagebrush for medical emergencies as serious as excessive bleeding after rattlesnake bites and hemorrhaging after difficult childbirths. Healers have apparently learned from both their mentors and from their own experiences which dosage levels their patients can tolerate.

Of course, Mitchell could not tell me whether particular Navajo patients he sees carry either the Naskapi or Mexico variants of serum albumin A. He simply knows that big sagebrush is potent, so potent that each patient must be carefully watched to determine a fitting dosage; it is not a one-size-fits-all medicine. For him, context is everything.

For some of the rest of us, coumarins may pose risks or benefits for reasons other than the serum albumin type we carry. Some of us harbor factor IX, a gene that increases coumarin sensitivity much as the Naskapi and Mexico albumin variants do. That means that we must be wary of exposure to warfarin and other substances that may trigger an adverse reaction in us. Others of us may have to deal with a polymorphism at the locus of the CYP2C9 gene, for a number of its alleles also increase coumarin sensitivity. Still other people carry a CYP2A6 gene for coumarin resistance, which allows them to rapidly bind any warfarin to which they are exposed, thereby stopping warfarin from limiting their blood-clotting capacity.

All in all, there are some sixty variations on this theme. That is, there are sixty different CYP genes in the P450 family. Many of these allelic variants can strongly influence how we metabolize some thirty prescribed and over-the-counter drugs, as well as countless foods and environmental chemicals.

Is it any wonder that the Navajo do not take lightly the culinary or medicinal use of sagebrush? Should a medicine man like Mitchell ever gather sagebrush roots, twigs, leaves, or pollen for you, you would need to reciprocate with a gift deemed of equal or exceeding value: an ancient arrowhead, a piece of turquoise jewelry, a hand-woven wool rug, or a supply of native foods.

~ Perhaps the very ambivalence of coumarins is why I shudder whenever I read how rapidly researchers in biotechnology, nutrigenomics, and pharmacogenetics are releasing new drugs and genetic therapy options to the public. It is now technologically possible to seek somatic gene therapy to reduce the potential hazards of ingesting coumarins. Anyone with enough money may soon choose to "rid" one's progeny of this problematic sensitivity to warfarin and other coumarins. How? By paying a biomedical genomics firm $1500 or more for a genetic profile that will determine if either member of a couple carries the Naskapi or Mexico variants of serum albumin A, or certain cytochrome genes in the P450 family. Once that is done, these parents could potentially ask a molecular biologist to deliver, via a carrier virus, a modified gene to either the developing fetus in the womb or to the newborn. Once the virus is injected into the so-called handicapped individual, it infects the cells that regulate the proteins causing coumarin hypersensitivity. Presto! If the therapy works, the child will no longer be "handicapped" or susceptible to the perils of coumarin consumption.

There you have it—it may soon be possible to produce a child less susceptible to coumarin toxicity, to carcinogenic activity, or to the ways warfarins reduce the clotting capacity of his or her blood. But now that same child may never be able to fully respond to the potency of sagebrush in a Navajo curing ritual, to its use in cleaning out intestinal parasites, to its ways of averting addiction to tobacco, or to its function as a sedative. The child's sensitivity to the world will be muted by gene therapy—fewer highs and fewer lows—and physicians will declare the child "normal." A so-called genetic disorder will be eliminated, but so will any adaptive value of the Naskapi or Mexico variants in certain environments. Thus, a hypothetical gene therapy could eliminate a health risk and paradoxically negate certain related health benefits at the same time.

There are, of course, other and perhaps more fruitful ways to respond to our newly enhanced knowledge of gene-food interactions in place of methods that wipe the genomic slate clean. Take, for instance, the extraordinary success of a nutritional strategy to reduce heart attacks and other diseases associated with clogged arteries. This strategy works, but only for those who are genetically vulnerable to one particular heart risk factor—elevated homocystine levels in the bloodstream. Homocystine is a toxic amino acid that increases in the bloodstream with the metabolic breakdown of protein, especially animal protein, in your diet. High homocystine levels account for one in every ten deaths from heart disease among men and nearly one in every twelve among women. Elevated homocystine levels in your bloodstream may also increase the risk of cancer and certain degenerative chronic diseases.

Once again, it was Arno Motulsky—the pioneer of nutritional ecogenetics discussed in chapter 3—who first noticed that an unusual gene-nutrition interaction was the underpinning of this story. This was significant because a rise in the bloodstream levels of homocystine was a risk factor for heart disease entirely independent of the susceptibilities that many populations suffer as a result of consuming too much of certain kinds of fats. As Motulsky later recalled to me, "In the search for genetic factors underlying premature heart disease, most attention has been given to genes affecting lipids. However, coronary and artery diseases often cannot be explained by genes affecting lipids. In the 1970s, [I noticed] an increasing number of reports suggesting that elevated levels of the amino acid homocystine were associated with various kinds of arteriosclerosis."

At first, scientists thought that there was some straightforward genetic link to high homocystine levels. They guessed that high levels originated with an inborn error in a recessive gene, a genetic error which led to a deficiency in an enzyme called cystathionine synthase. The scientists guessed that heterozygotes with only one copy of the recessive gene were the individuals most likely to have the condition they dubbed *homocystinuria*. Those who have this condition were found more likely to suffer from arterial or heart disease.

But as genetic studies became more refined, scientists discovered that no heterozygotes for this gene could be found at all among hundreds of Dutch and Irish sufferers of premature heart disease. This perplexed the biomedical community for a number of years until other scientists noticed that a different gene was influencing high homocystine levels. It was one with which nutritionists and geneticists were already familiar, for it played a key role in folic acid metabolism.

The name *folic acid* comes from *foliage*, and refers to one of those unnumbered B vitamins found in leafy greens, beans, and some fruit juices. The second gene that influences homocystine levels was already known to foster the efficient use of folic acid and vitamin B12 to ward off pernicious anemia. The gene also allows our bodies to utilize folic acid as a precursor for a large family of folates that serve as coenzymes for energy transfers within the body. Furthermore, this gene is needed for the production of an enzyme nicknamed MTHFR—methylene tetrahydrofolate reductase. Those who lack this MTHFR enzyme due to a homozygous recessive trait are three times more likely to suffer from arterial clogging and heart disease than those who are heterozygous or homozygous dominants.

This sounds as though those with severe MTHFR deficiency have a sealed fate, but that is not the case. The same team of researchers that discovered the link between MTHFR deficiency and coronary disease soon made an even more profound discovery: increased consumption of folic acid could markedly reduce the probability of arterial clogging and heart attacks.

When the implications of that discovery dawned on nutritionists, they quickly mobilized an effort to see whether dietary supplementation of folic acid could demonstrably decrease homocystine levels in vulnerable populations. It was effective, but *only* for homozygous recessive individuals! It seemed as though a solution to high homocystine levels was close at hand. But as Professor Motulsky observed, rather than being a quick genetic fix, the solution would more likely take the form of a nutritional intervention that would benefit some but not all comers: "This finding is an important example of nutritional/genetic interactions, in that it shows that homocystine elevations occurred only when folic acid nutrition was less than optimal—that is, among persons in the lower one-half of the distribution of plasma folic acid levels."

In 1995, Motulsky and his colleagues suggested that because there was a simple, inexpensive screening technique for the MTHFR polymorphism, a screen-and-treat strategy could effectively extend the lives of many individuals who would otherwise die prematurely due to low folic acid intake. The alternative, he suggested, would be to treat the entire population by fortifying key foods with 350–400 micrograms of folic acid per day.

Medical researchers determined that the screen-and-treat strategy would be more cost effective than universal supplementation, and certainly less expensive than surgical interventions to deal with heart disease after the fact. With cold calculations, researchers claimed that screening followed by folic acid and vitamin B12 fortification of foods would cost $2.1 billion in the United States and lead to 122,000 years of life "saved." In contrast, the "treat-all" strategy would cost $5.5 billion—more than twice as much—but save just 4,000 more years of life, or 126,000 life years total. However, if pregnant mothers ate these fortified foods, it would also reduce spina bifida and other neural-tube defects that cause tragic fetal abnormalities.

The folic acid link to spina bifida suddenly changed the gamble. With impetus to reduce both heart disease and birth defects, the U.S. Food and Drug Administration opted for the treat-all strategy. It decided to require that all enriched cereal-grain products be fortified with folic acid by January 1998. Aiming to increase the average person's intake just 100 micrograms per day, folic acid was added to different products in amounts ranging from 95 micrograms to 309 micrograms per 100 grams of product. Remarkably, within a matter of a few years, the average consumer was obtaining 190 more micrograms of folic acid per day without taking any vitamin and mineral supplements, almost twice the levels that the FDA had hoped to achieve. Those who also took vitamin and mineral supplements containing folic acid increased their average intake to 219 micrograms per day.

The effects on reducing the risk of heart disease have been astonishing. While about half of all consumers failed to receive the recommended amount of folic acid before gaining access to fortified foods, only 7 percent failed to achieve the recommended intakes after fortified foods became available. Increasing folic acid intake to around 200 micrograms a day dropped homocystine levels significantly. As homocystine levels went down, the numbers of heart and artery diseases were reduced by at least 13,500 per year and perhaps by as much as 50,000 per year in the United States. In less than five years, the fortification effort had essentially increased folic acid intake among MTHFR-deficient individuals to the degree that they were no longer any more susceptible to heart disease than others were. It has become clear that a rather expensive somatic gene therapy is not necessary to improve the health of those with homocystine deficiencies. Instead, attention to nutrition, and specifically, to gene-nutrient interactions, can be enough to do the trick.

There are, of course, other scientists who would claim that a genetic fix—not for humans, but for our grain crops—could ultimately be the most cost-effective and elegant solution to our need for folic acid and for vitamins such as B12. Why mess with people's genes, they argue, if we can more expediently tinker with the genes of foods in ways that can make up for our genetic vulnerabilities? Let's see what they mean.

While most green leafy vegetables are naturally rich in folic acid, grains are not. It is feasible to implant genes from leafy greens into cereals to increase folic acid. These inserted genes could increase the production of folic acid in a cultivated grass, or they may enable a shunt in the grasses' metabolic pathways that results in more folic acid being retained in the cereal grain itself. By manipulating just six cereal crops that provide humankind with the bulk of our calories—rice, barley, wheat, sorghum, millet, and maize—couldn't we get all the folic acid we need into our grains even before they are ground?

The answer to that hypothetical scenario is this: the simplest solution to folic acid deficiency need not be found in a biotech laboratory, for it is already outside our back door. As Bill McKibben has proclaimed in *Enough*, "What you need is not miracles from Monsanto; what you need is a diet rich in local greens!"

Although that recommendation may be a little too cryptic for the uninitiated, McKibben is suggesting that there is really no need for biotechnology firms to manipulate either our genes or the genes of our food crops if we maintain our consumption of fresh, green leafy vegetables. And that is what MTHFR deficiency may have been saying—in an evolutionary sense—to countless generations of residents of northern Europe anyway. Because of their short growing season, the Scandinavians, Greenlanders, Danish, Dutch, Scotch, British, and Irish have traditionally had an easier time of keeping meat in their diet during all seasons than leafy greens. Protein- and fat-rich meats, while nutritious, elevate homocystine levels and increase the risk of heart disease and cancer. This is particularly true for MTHFR-deficient northerners *unless* frequent consumption of leafy green vegetables brings homocystine levels back down.

The farther north you live, the tougher it is to get such greens—the Inuit of the Arctic Circle must obtain them from the lichen "stomach salads" removed from caribou innards—but the more critical such greens are to your overall health. Why? Greens not only provide folic acid; they are also among the best sources of several vitamins, minerals, and dietary fibers. They not only contain a variety of antioxidants and immune system boosters, they provide protection against scurvy and other maladies emerging from diets chronically limited in fresh vegetables and fruits. In short, those with MTHFR deficiency who did *not* develop culinary traditions that maximized fresh greens suffer from high homocystine levels that trigger heart disease; such people are also vulnerable to a plethora of other health problems as well. Natural selection clearly favored the MTHFR-deficient who were green eaters in northern climes; in Mediterranean and tropical countries, where greens where available year-round, the genes that elevate homocystine levels are far less frequent.

~ I thought about this paradox during the only week of my life that I have spent in the British Isles, where the fashion of the last several centuries appears to have favored high-cholesterol delights over beans and greens—diets that include steak and kidney pies, fish and chips, or calf brains with black butter. As a result, the prevalence of heart disease in England remains wickedly high.

As I wandered from pub to pub, I was constantly struck by the miracle of the British surviving so long on a diet rich in meat protein and fat, fiber-depleted cereal flour and roots starches, and ales and whiskies, especially since the latter interfere with both folic acid absorption and metabolism. I did not encounter a fresh, healthy looking, green leafy vegetable all week long.

When I cornered a British friend who had recently returned from holiday in Morocco, after stints in Mexico and Malaysia, I asked if he could provide me with the recipe of a truly traditional food from his family. I hoped that the nostalgia and patriotism he might be feeling having returned home would lead him to convince me that the limited UK diet I had thus far encountered was some fluke. He took another swallow from his mug of stout, grinned, and uttered something that sounded like "Toad Inner Hole." Ah, I thought to myself, perhaps his family had eaten amphibians now and then as a delicacy.

"Toad Inner Hole?"

"Noooo. Toad *In the* Hole!"

"What's in it?"

"In the hole? The bloody toad!"

"No, I mean in the recipe!"

He took another draught of his ale, and sighed. "Well, it's very simple, but absolutely traditional here: four or five sausages, fried in pork fat, and covered in a layer of Yorkshire pudding and butter so that only the butts of the sausage stick up."

"Like toads in a toxic swamp!" I blurted out. "Meat, grease, and flour, fried in more grease? It's a wonder that everyone born and bred in England hasn't already had a heart attack and died!"

He coughed, and then shrugged. "Well, as a matter of fact, perhaps that's why so many of us live as ex-pats, spending most of our adult lives doing work abroad, surviving on tropical fruits, curries, and *molés*. . . . Yes, I believe you're right, lad. We'd have all been dead by now if we had stuck it out our entire lives in English pubs. If not dead, then continuously drunk. That *is* the reason why the British are so blessedly colonial: we migrate to the four corners of the Earth as soon as we are able simply to get away from our own bloody food." My friend was kidding, of course: we cannot so easily leave behind our ancestral food preferences. They are embedded in our bodies and our minds, just as deeply as the toad is embedded in a pool of highly rendered fat.

CHAPTER SEVEN

# *Rooting Out the Causes of Disease*

## Why Diabetes Is So Common Among Desert Dwellers

FROM THE LAND OF THE NAVAJO, let us go southward into Mexico once again, to a coastal community of another indigenous people. Although genetically unrelated, the Navajo of the United States and the Seri of Mexico share a problem that has both a genetic and a nutritional component: adult-onset diabetes. This nutrition-related disease is one of the three top causes of death among these two Native American groups and among many other indigenous communities as well. Ironically, a half century ago, its presence as a health risk was so minor in these communities that more Indians were dying each year of accidental snake bite than of diabetes. To understand why that change occurred, and what it means for all of us, we must listen not just to epidemiologists, but to the native peoples themselves.

It was in a small, run-down health clinic on a beach of Mexico's Sea of Cortés that an Indian elder gave me a memorable lesson about gene-food interactions. It was a lesson nested in place—the hot desert coastline studded with giant cactus; that particular Indian village, where people cooked most of their food on small campfires in the sandy spaces between shabby government-built houses; and in that clinic, with no windows and no equipment, so rarely frequented by a doctor that we had planted a garden of healing herbs around it in case there was ever a medical emergency. It was in this place that Seri Indian Alfredo López Blanco challenged me—and Western-trained scientists in general—to pay protracted attention to diet change and its role in disease.

I had accompanied my wife Laurie Monti, nurse-practitioner turned medical anthropologist, who was screening Seri families for adult-onset diabetes. The disease was already running rampant through neighboring tribes, but because the Seri are the last culture in Mexico to have retained hunting, fishing, and foraging traditions instead of adopting agriculture, there was some hope that they could stave it off. Only a few of the some 650 tribal members had ever been screened for the noninsulin-dependent form of diabetes, and that smaller, earlier sample had suggested that only 8 percent of the tribe suffered from chronically high blood-sugar levels and low insulin sensitivity.

While Laurie was screening Seri families in the sole office that contained any semblance of sanitary surfaces, I was in the "waiting room"—a sort of stripped-down echo chamber full of barking dogs and crying babies—trying to interview the elders of each family about their genealogical histories. I was attempting to ascertain whether the genetic susceptibility to diabetes of individuals with 100 percent Seri ancestry might be different from those who claimed that some of their ancestors came from among the neighboring Pima and Papago (O'odham) tribes in Arizona, the ethnic populations reputed to have the highest incidence of diabetes in the world.

Alfredo López Blanco returned to the waiting room after Laurie confirmed that his blood-sugar levels were unusually high. Alfredo, who had worked as a fishermen since he was a boy, had late in life become boatman and guide for marine and island biologists. In his late sixties, Alfredo often taught younger Seri about the days when their people had subsisted on seafood, wild game, and desert plants like cactus fruit and mesquite pods. He was keenly aware of the traditional diet of his own people, and of his neighbors as well. When he sat down with me, I asked him if any of his forefathers happened to be from other tribes. He answered that one of his great-grandmothers was from a Papago-Pima community.

"But *Hant Coáaxoj*," he called me by my Seri nickname, Horned Lizard, "I have a question for you. What does that have to do to my diabetes?"

"Well, I'm not yet sure that it does. But here's why I'm asking. The Pima and Papago suffer from diabetes more than any other tribes. It might be in their blood," I conjectured, groping for a way to explain the concept of *genetic predisposition* to a person whose native language does not contain the exact concept of "genes." "If people have Pima blood in them, maybe they are more prone to diabetes."

"*Hant Coáaxoj*," he said dryly, "sometimes you scientists don't know much history. If diabetes is in their blood—or for that matter, in our blood—why did their grandparents not have it? Why were the old-time Pima and Papago who I knew skinny and healthy? It is a change in the diet, not their blood. They are no longer eating the bighorn sheep, mule deer, desert tortoise, cactus fruit, and mesquite pods. *Pan Bimbo* bread, Coke, sandwiches, and *chicharrones* are the problem!"

The old man—whose sister died within a year of that conversation due to circulatory complications from her own diabetes—was pretty much right on the mark. Or at least that is what Laurie's interpretation of her screening and my genealogical interviews later showed. Diabetes, aggravated by diet change, was clearly on the rise among the Seri, with more than 27 percent of the adults screened by Laurie showing impaired glucose tolerance. But there were also interesting differences between the village where Alfredo lived, Punta Chueca, and the more remote Seri village to the north, Desemboque, where Western foods and other signs of acculturation were much less prominent. While diabetes prevalence in Desemboque had only recently reached 20 percent of the adults in Laurie's sample, it exceeded 40 percent in Punta Chueca.

Other public-health surveys of the Seri suggested why this might be the case. Punta Chueca's residents had easier access to fast-food restaurants and minimarts than did Desemboque dwellers. Roughly 15 percent more of Punta Chueca's residents consumed groceries purchased in nearby Mexican towns on a daily basis, rather than relying more heavily on native foods from the desert and sea. The people of Punta Chueca consumed significantly more store-bought fats (such as lard), alcohol, and cigarettes.

When data from both villages were pooled, Seri individuals with some Papago-Pima ancestry did not show up as suffering from diabetes any more than those with 100 percent Seri ancestry. And yet, comparing the villages, there was one telling difference: those with Papago-Pima ancestry who ate more acculturated, modernized diets in Punta Chueca had the highest probability of the disease. As long as the Desemboque dwellers with Papago-Pima blood remained close to their traditional diet, diabetes among them was held more in check.

This trend held even though traditional Seri individuals in Desemboque appeared to weigh somewhat more than their counterparts in Punta Chueca. This suggests that it may not be the sheer quantity of food metabolized that triggers diabetes as much as the qualities of the foods the Seri now eat—especially the kinds of fats and carbohydrates regularly consumed.

This key distinction has slipped past the U.S. National Institutes of Health Indian Diabetes Project in the Sonoran Desert, which for nearly four decades has spent hundreds of millions of dollars trying to identify the underlying cause of the diabetes epidemic among the Pima and other indigenous communities. Its scientists and educators have all but ignored qualitative differences between Native American diets, preferring to seek a quick genetic fix to everyone's problem at the same time. Several years ago, *New Yorker* writer Malcolm Gladwell called it the "Pima paradox": "All told, the collaboration between the NIH and the Pima is one of the most fruitful relationships in modern medical science—with one fateful exception. After thirty-five years, no one has had any success in helping the Pima lose weight [and control diabetes]. For all the prodding and poking, the hundreds of research papers describing their bodily processes, and the determined efforts of health workers, year after year the tribe grows fatter."

At most, the NIH epidemiologists have quantified how the contemporary Pima and their Indian neighbors eat more fast foods than ever before, especially ones detrimentally high in animal fats and simple sugars. But what the NIH has failed to discuss with Native Americans are the countless studies, including my own collaborations with nutritionists, that demonstrate how traditional diets of desert peoples formerly *protected* them from diabetes and other life-threatening afflictions now known as Syndrome X. This Syndrome X is not some sinister new disease, but rather a cluster of conditions that, when expressed together, may reflect a predisposition to diabetes, hypertension, and heart disease. The term—first coined by members of a Stanford University biomedical team—describes a cluster of symptoms, including high blood pressure, high triglycerides, decreased HDL ("good" cholesterol), and obesity. These symptoms tend to appear together in some individuals, increasing their risk for both diabetes and heart disease. And of course, all of these symptoms are influenced by diet, but what kind of diet most effectively reduces their expression was something that I seemed more interested in than anyone at the NIH or at Stanford.

~ In the 1980s, I began to collect traditionally prepared desert foods for nutritional analysis by Chuck Weber and Jim Berry in their University of Arizona Nutrition and Food Sciences lab, and for glycemic analysis by Jennie Brand-Miller and her colleagues who had already done similar work analyzing the desert foods traditionally consumed by Australian aborigines. By glycemic analysis, I refer to a simple finger-prick test for blood-sugar and insulin levels done as soon as a particular food is eaten, and every half hour afterwards; the test determines whether the food in question causes blood-sugar levels to rapidly spike after its ingestion, thereby causing pancreatic stress and asynchronies with insulin production.

Jennie Brand-Miller, a good friend as well as colleague, determined with her students that native desert foods—desert legumes, cacti, and acorns in particular—were so slowly digested and absorbed that blood-sugar levels remained in sync with insulin production, without any adverse health effects generated. Jennie called these native edible plants "slow-release foods" to contrast them with spike-inducing fast foods such as potato chips, sponge cakes, ice cream, and fry breads. The fast foods had glycemic values two to four times higher than the native desert foods, whose slow-release qualities Weber and Berry had shown to be derived from the foods' higher content of soluble fiber, tannins, and complex carbohydrates.

Jennie had found the same trend when comparing Western fast foods with the native desert foods that Aussies call "bush tuckers"—the mainstays of aboriginal diets up until a half century ago, before which diabetes was virtually absent in indigenous communities of Australia. As with the desert tribes of North America, once these protective foods were displaced from aboriginal diets, the incidence of diabetes skyrocketed.

Back on an autumn night in 1985, Jennie and I were sipping prickly pear punch, having spent the day comparing the qualities of Australian and American desert foods. I could see that she was brewing over some large question, and she finally teased it out.

"Gary, I've wondered if there might be some explanation [for why desert peoples are vulnerable to diabetes, other than what the NIH promotes], one that you as a desert plant ecologist might help me figure out. I don't know if I'm framing this question precisely enough, but let me give it a try: is there something that helps a number of desert plants adapt to arid conditions which might help control blood-sugar and insulin levels in the humans that consume them?"

"*What*?" I blurted out. "Could you say that again?" Much later, I thought of a famous comment about the heart of science: "Ask an impertinent question and you are on your way to a pertinent answer."

Jennie laughed, aware that she was asking a question far too complex to consider in the midst of the frivolity of a dinner party. "Oh, that's OK," she said quietly. "I just wondered if desert plants from around the world could have evolved the same protective mechanism against drought that somehow. . . . "

"Oh, I think I get it now, some kind of convergent evolution," I said. "If the same drought-adapted chemical substances show up in plants from various deserts that are scattered around the world, perhaps these substances formerly protected the people who consumed them from the risk of diabetes. . . . " Then, once diets changed, the desert peoples who once had the best dietary protection from diabetes suddenly had their genetic susceptibility expressed!

Although Jennie posed it in passing, I could not forget her impertinent question, not that night, not that week, and not for a long time. Friends like Gabriel, as well as Alfredo López's sister, Eva, had died of diabetes, but they still inhabited my memory. I mused over Jennie's question whenever I was out studying plants in the desert, and I brought it to the attention of some physiological ecologists who had a far deeper understanding of plant adaptations to drought than I did. They reminded me that desert plants and animals adapted to drought conditions by many different anatomical, physiological, and chemical means, and that there was probably not a single protective substance found in all arid-adapted biota.

In other words, the flora and the fauna from different deserts emphasized distinctive sets of these adaptive strategies. It was simply too much for these ecologists to imagine that a cactus from American deserts and a *wichitty* grub from the Australian outback might all share some dietary chemical that controlled diabetes among the Pima, Papago, and Seri of American deserts as well as among the Warlpiri and Pinkjanjara of Australian deserts.

Still, Jennie's question was rooted in a valid observation: there was an apparent correlation between the extraordinarily high susceptibility of diabetes among desert peoples and the quantity of drought-adapted plants in their diets. If some aboriginal cultures had subsisted on drought-adapted plants and associated wildlife for upwards of 40,000 years, was it not plausible that these people's metabolisms had adapted to the prevailing substances in these foodstuffs? And if, within the last fifty years, the prevalence of these foodstuffs had declined precipitously in their diets, was it not just as plausible that they had suddenly become susceptible to nutrition-related diseases because they had *lost their protection*? The question to pursue, then, was what dietary chemicals—nutrients or even antinutritional factors—might be more common in drought-adapted plants than those occurring in wetter environments?

With the help of ecophysiologist Suzanne Morse, I tried to imagine how water loss from a plant's tissue was slowed by the adaptations developed by a desert-dwelling organism to deal with scant and unpredictable rainfall. At the time, I was involved in a number of field evaluations of drought tolerance in desert legumes, cacti, and century plants. I soon learned that prickly pear cactus pads contain *extracellular mucilage*, that is, gooey globs of soluble fiber that holds onto water longer and stronger than the moisture held within photosynthesizing cells. If a cactus is terribly stressed by drought, it may shut down its photosynthetic apparatus, shut its stomatal pores, and shed most of its root mass, going "dormant" until rain returns. But if stress is not so severe, the cactus will instead gradually shunt the moisture in its extracellular mucilage into photosynthetically active but water-limited cells, thereby slowing the plant's total water loss while keeping active tissues turgid.

In explaining this concept—called "leaf capacitance"—to me, Morse offered me a parallel to slow (sugar) release foods: slow (water) release plant tissues. The very mucilage and pectin that slow down the digestion and absorption of sugars in our guts are produced by prickly pears to slow water loss during times of drought. And prickly pear, it turns out, has been among the most effective slow-release foods in terms of helping diabetes-prone native peoples slow the rise in their blood-glucose levels after a sugar-rich meal. In fact, it was among the first foods native to the Americas demonstrated to lower the blood glucose and cholesterol of indigenous people susceptible to diabetes. As Morse and I followed up on that research, we documented that most of the twenty-two species of cacti traditionally used by the Seri have the same slow-release qualities and are available along the desert coast much of the year.

Soon, Jennie Brand-Miller, in Sydney, and Boyd Swinburn, an endocrinologist from New Zealand, gave me greater insight into how slow-release foods differ from conventional foodstuffs in they way they are digested and absorbed. As I read reports about the "low gastric motility" of slow-release foods, I began to imagine how these foods make a viscous, gooey mass in our bellies. Even when our digestive juices cleave them into simpler sugars, the sugars have a tough time moving through the goo to reach the linings of our guts, to be absorbed and then transported to where they fuel our cells.

Here then, in the prickly pear—one of the food plants in the Americas with the greatest antiquity of use—was the convergence that Jennie had been seeking: the existence of slow-water-release mucilage in cactus pads and fruit explained why desert food plants were likely to produce slow-sugar-release foods. Five years after our conversation over prickly pear punch, I found a potential answer in the very plant

Jennie and I had been consuming at the time she asked her impertinent question! The trouble was, prickly pear and other cacti are not native to Australian deserts; I began to investigate if there were plants in other deserts that also contained slow-water-release mucilages.

I soon learned that cacti are not special cases that occur only in the diets of desert-dwelling Native Americans; there are dozens of other plants in both American and Australian deserts that have similar slow-sugar-release/slow-water-loss qualities, albeit with different morphologies and different chemical mechanisms. Given that desert peoples have been exposed to such plants for upwards of 10,000 years—more than 40,000 years in Australia—is there any evidence that these people's metabolisms have adapted over time to the presence of these protective foods?

With regard to the Seri, the only general genetic survey comparing them to neighboring agricultural tribes indicates that the Seri exhibit "several micro-polymorphisms [that] may be important in conferring a biological advantage" in their desert coastal homeland. The study claimed that "these may emphasize the relevance of interactions between genes and environment," for Seri hunter-gatherers express several alleles not found in more agriculture-dependent U.S. and Mexican indigenous peoples (Infante et al. 1999).

But do long-time hunter-gatherers with such polymorphisms respond to certain desert and marine foods differently than other people do? The answer can be found in research that Jennie and colleagues have done contrasting various ethnic populations' responses to foods common to one group's traditional diet, but not the other's. As Jennie and her fellow researcher Anne Thorburn have explained,

> the aim of [our] next series of experiments was to compare the responses of healthy Aboriginal and Caucasian subjects to two foods, one a slow release Aboriginal bush food—bush potato (*Ipomoea costata*)—and the other a fast release Western food—[the domesticated] potato (*Solanum tuberosum*). Both Aborigines and Caucasians were found to produce lower plasma insulin responses to the slow release bush food than to the fast release Western food. But the differences were more marked in Aborigines, with the areas under the glucose and insulin curves being one-third smaller after bush potato than potato" (Brand-Miller and Thorburn 1987).

In other words, the Aborigines were protected from diabetic-inducing pancreatic stress by a bush food that their metabolisms had genetically adapted to over 40,000 years. Caucasians, with hardly any exposure to this or similar bush foods since colonizing Australia, did not experience such marked benefits.

When many scientists learn of these differences, they recall the theory of a *thrifty gene* that indigenous hunter-gatherers are presumed to maintain as an adaptation to a feast-or-famine existence, and they attribute the differences in insulin response to that gene. As originally hypothesized by James Neel in 1962, hunter-gatherers were likely to exhibit a thrifty genotype that was a vestigial survival mechanism from eras during which they suffered from irregular food availability. "During the first 99 percent or more of man's life on earth while he existed as a hunter-gatherer," Neel wrote, "it was often feast or famine. Periods of gorging alternated with periods of greatly reduced food intake" (Neel 1962).

Neel persuasively argued that repeated cycles of feast and famine over the course of human evolution had selected for a genotype that promoted excessive weight gain during times of food abundance and gradual weight loss of those "reserves" during times of drought. Neel focused on food quantity—the evenness of calories over time—and not food quality, arguing that when former hunter-gatherers were assured regular food quantities over time, the previously adaptive genetic predisposition to weight gain became maladaptive.

However, the only early NIH attempt to characterize the diets of Pima women with traditional versus acculturated (modern) lifestyles found insignificant differences between the calorie amounts consumed by the two groups, nor was there much difference when both groups' diets were compared to what surrounding Anglo populations ate. In other words, despite Neel's hypothesis, food quantity alone did not account for the rise in diabetes among acculturated Pima Indian women.

Nevertheless, Neel's argument has been cited by hundreds of scientific papers on diabetes and other diseases and has reached millions of other readers through "popular science" magazine essays written by such science-literate writers as the *New Yorker*'s Malcolm Gladwell, *Harper's* Greg Cristner, *Outside*'s David Quammen, and *Natural History*'s Jared Diamond. What's more, Neel's hypothesis essentially drove the first thirty-five years of research at the NIH Indian Diabetes Project in Phoenix, Arizona, whose director and staff set their sights on becoming the first to discover *the* thrifty gene. Hundreds of millions of research dollars later, it is clear that their focus on a single gene and on sheer food *quantity* has blinded researchers to a variety of gene-food-culture interactions that may trigger or prevent diabetes.

Thirty-six years after proposing his famous hypothesis, Neel himself conceded that "the term 'thrifty genotype' has [already] served its purpose, overtaken by the growing complexity of modern genetic medicine," adding that while type 2 diabetes may still be "a multifactorial or oligogenic trait, the enormous range of individual or group socioeconomic circumstances in industrialized nations badly interferes with an estimate of genetic susceptibilities" (Neel 1998).

Neel's colleagues in biomedical research are much more direct in their assertion that there is no single thrifty gene that confers susceptibility to type 2 diabetes among all ethnic populations, or even among all hunter-gatherers. Assessing the recent identification of several genes that heighten or trigger diabetes, geneticist Alan Shuldiner of the University of Maryland School of Medicine told *Science News*, "I expect there would be dozens of diabetes-susceptibility genes [and that] specific combinations of these genes will identify risk" (Seppa 2002).

What these genes actually do is also different from what Neel and other proponents of the thrifty genotype suspected they would do. When the NIH worked to determine whether *the* thrifty gene they had identified in the Pima was actually a gene for insulin resistance—

which causes reduced metabolic sensitivity to sugar loads—researchers found this gene's true function to be weight maintenance and *not* weight gain.

As molecular biologist Morris White of the Joslin Diabetes Center recently concluded in the pages of *Science*, "We used to think type 2 diabetes was an insulin receptor problem, and it's not. We used to think it was solely a problem of insulin resistance, and it's not. We used to think that muscle and fat were the primary tissues involved, and they are not. Nearly every feature of this disease that we thought was true 10 years ago turned out to be wrong" (White 2000).

Once again, it was my friend Jennie Brand-Miller who hammered the coffin closed on the thrifty gene hypothesis by refuting its very underpinnings—that famines were more frequent among hunter-gatherers than among agriculturists, leading to the former's extraordinary capacity to accumulate fat reserves. In scanning the historic anthropological literature on periodic famine and starvation among various ethnic groups, Jennie and her colleagues found scant evidence that hunter-gatherers suffered from these stresses anywhere near as frequently as agriculturalists did. In fact, periodic starvation and widespread famines increased in frequency less than 10,000 years ago, after various ethnic groups became fully dependent on agricultural yields. In particular, Jennie noted, since Caucasians living in Europe have repeatedly suffered from famines in historic times, they ought to be predisposed to insulin resistance and diabetes if Neel's hypothesis is correct. And yet, Caucasians are one of the few groups that do not exhibit much insulin resistance or heightened susceptibility to type 2 diabetes when they consume modern industrialized agricultural diets.

"The challenge," Jennie and her colleagues argue, "is to explain how Europeans came to have a low prevalence and low susceptibility to adult-onset diabetes . . . " (Cordain et al. 2000). Indeed, Europe harbors most of the world's ethnic populations who have *not* suffered dramatic rises in this nutrition-related disease since 1950.

At an international workshop that Jennie and I hosted at Kims Toowoon Bay on the coast of New South Wales in May of 1993, we elucidated four factors that could explain why individuals of European descent appear to be less vulnerable to Syndrome X maladies—including diabetes—than do ethnic populations that have adopted agricultural and industrial economies more recently. With colleagues from four countries, including Australian Aborigines and Native Americans, we identified that the incidence of diabetes rapidly increases under the following four circumstances.

First, when an ethnic population shifts to an agricultural diet and abandons a diverse cornucopia of wild foods, its members lose many secondary plant compounds that formerly protected them from impaired glucose tolerance. This is particularly true for populations that have coevolved with a certain set of wild foods over millennia, ones that are rich in antioxidants.

Second, when the remaining beneficial compounds in traditional crops and free-ranging livestock are selected out of a people's diet through breeding and restricted livestock management practices, their diet is further depleted of protective factors. For instance, modern bean cultivars have been bred to contain less soluble fiber, and livestock raised on cereal grains under feedlot conditions lack omega-3 fatty acids.

Third, the industrial revolution that began in Europe in the seventeenth century changed the quality of carbohydrates in staple foods by milling away most of the fiber in them. High-speed roller mills now grind grains into easily digested and rapidly absorbed cereals and flours, which results in blood-sugar and insulin responses two to three times higher than those reported from whole grains or coarse-milled products like bulgur wheat.

Fourth, the last fifty years of highly industrialized foods has introduced additives such as trans-fatty acids, fiber-depleted gelatinous starches, and sugary syrups, which ensure that most fast foods are truly *fast-release* foods. Jennie estimates that the typical fast-food meal raises blood-sugar and insulin levels three times higher than humans ever experienced during preagricultural periods in our evolution. Combined with the trend toward oversize servings of convenience foods and a more sedentary lifestyle, the dominance of fast foods in modern diets has made contemporary humans less fit than ever.

Although nearly all ethnic populations have come to suffer from fast foods over the last quarter century, the other changes took place in European societies over thousands of years. Whereas the genetic constituency of European peoples may have slowly shifted with these technological and agricultural changes as they emerged, the Seri and Warlpiri have had less than fifty years to accommodate these changes, and their genes are not in sync with them. Significant adaptation through evolutionary processes to new diets rarely occurs over the course of two to three generations.

And yet, most people now living in the world fall somewhere between the French and German farmers on the one hand, and the Seri and Warlpiri hunter-gatherers on the other. The majority of traditional diets have historically been more like the Pima and Papago in the Arizona deserts, where perhaps 60 percent of foods were harvested from domesticated crops in wet years while the rest came from wild and weedy plants and free-ranging game or fish. In dry years, the Papago-Pima diet shifted more toward the reliable harvests of drought-tolerant wild perennials.

While details richly vary around the world—from coastal habitats where fish were once abundant to rain forests where birds and root crops proliferated—most indigenous peoples in developing countries have maintained, until recently, a healthy mix of wild foods and diverse cultivated crops. Today, following dramatic economic shifts that have favored a few cereal grains and livestock production for export over mixed cropping, the bulk of the world's population has been left vulnerable to diabetes. One recent reckoning suggests that upwards of 200 million people are now susceptible to diabetes and the other killers associated with Syndrome X. This is not the exception among the diverse peoples of the world; it is a pathology that has become the norm.

But while fast foods lead to rapid deterioration of healthy carbohydrate metabolism in most people—with or without the existence of a thrifty gene—a return to the traditional foods of one's own ancestry leads to rapid recovery. This is what New Zealand endocrinologist Boyd Swinburn found when he asked me to help him reconstruct a semblance of the nineteenth-century dietary regime for the Pima and Papago. Swinburn wanted to compare the effects of a traditional versus a fast-food diet, both consisting of the same number of calories and percentages of carbohydrate and fats.

When twenty-two Pima Indians in his study were exposed to the fast-food diet, their insulin metabolism deteriorated enough to trigger diabetic stress without the need to conjure up any other explanation to explain it. Yet when the same individuals were placed on the traditional diet rich in soluble fiber and other secondary plant compounds, their insulin sensitivity and glucose tolerance improved. Swinburn and his coworkers concluded that "the influence of Westernization on the prevalence of diabetes may in part be due to changes in dietary composition [as opposed to food quantity]" (Swinburn et al. 1993).

I followed Swinburn's clinical study with a demonstration project at the National Institute for Fitness outside St. George, Utah, where eight Pima, Papago, Hopi, and Southern Paiute friends suffering from diabetes came together for ten days of all-you-can-eat slow-release foods and outdoor exercise. Within ten days, their weight and their blood-sugar levels had been dramatically reduced, and everyone felt healthier. The changes began so immediately that several participants had to seek medical advice to figure out how to reduce the hypoglycemic medications they had been self-administering for years.

In yet another example, in what may be one of the most dramatic gains in health conditions ever witnessed in a short period of time, Kieran O'Dea documented the marked improvement in diabetic Australian aborigines after they reverted for a month to a nomadic foraging lifestyle in western Australia. Even though study subjects "poached" several free-ranging cows as part of their meat consumption, their diet primarily consisted of bush foods that their ancestors had long eaten. The aboriginal participants moved frequently to take advantage of hunting and plant-gathering opportunities, and they lost considerable weight while doing so.

Their consumption of calories from macronutrients was 54 percent protein, about 20 percent plant carbohydrates, and 26 percent fat. These proportions had a dramatic effect on lowering blood-sugar levels and increasing insulin sensitivity. While some critics have conjectured that their insulin sensitivity, glucose tolerance, and cholesterol levels improved merely because of the subjects' weight loss, others have pointed out that the ratio of macronutrients they consumed certainly did not worsen their condition. While not necessarily optimal for all ethnic populations, a diet with this mixture of macronutrients clearly brought health benefits to the Australian desert dwellers that participated.

~ Inspired by O'Dea's collaboration with indigenous sufferers of diabetes, I organized a similar moveable feast in the spring of 1999, engaging more than twenty Seri, Papago, and Pima individuals who also suffered from diabetes. We walked 220 miles through the Sonoran Desert during a twelve-day pilgrimage, fueled only by native slow-release foods and beverages. Although we did not measure our blood-sugar and insulin levels each day to compare our health status before and after our journey, we took note of something perhaps far more significant: the native foods we ate were considered by all the participants to be nutritious, satisfying, and filling enough to sustain our arduous pilgrimage. These foods enabled us to hike across rugged terrain for ten hours a day, followed by another hour or two of celebratory dancing. Our collective effort made us more deeply aware that our own energy levels could be sustained for hours by slow-release foods. At the same time, we took a good hard look at the health of our neighbors and of the land itself. The pilgrimage allowed us to clearly see for the first time all the damage that had been done to our homeland and its food system, damage that was echoed in our very own bodies.

There was something else going on among my Native American companions during that walk. The Seri, Papago, and Pima pilgrims frequently expressed that their cultural pride, spiritual identity, and sense of curiosity were being renewed. And so, a return to a more traditional diet of their ancestral foods was not merely some trip to fantasy land for nostalgia's sake; it provided them with a deep motivation for improving their own health by blending modern and traditional medical knowledge in a way that made them feel *whole*. They were not eating native slow-release foods merely to benefit a single gene—thrifty or not. Instead, they were communing to keep their entire bodies, their entire communities, and the entire Earth healthy.

Yes, genes matter, but diverse diets and exercise patterns matter just as much. And when the positive interaction among all three of these factors is reinforced by strong cultural traditions, our physical health improves, as does our determination to keep it that way. The Native American folks I walked with on that pilgrimage have redoubled their commitments to keep their traditional slow-release foods accessible in their communities; they serve them at village feasts and at wakes honoring those who have succumbed to the complications of diabetes for lack of earlier access to these foods. When the persistence of traditional foods is more widely recognized as a source of both cultural pride and as an aid to physical survival and well-being, I doubt that many Native American communities will abandon what many of them feel to be a true gift from their Creator.

CHAPTER EIGHT

# *Reconnecting the Health of the People with the Health of the Land*

## How Hawaiians Are Curing Themselves

THE LAST LEG of our odyssey together takes us to Hawaii, where many strands from the previous stories intertwine to remind us that we are not simply talking about genes or genotypes; we are talking about lives—the vibrant lives of remarkable individuals and diverse cultural communities—and the choice to work toward saving them or to fatalistically stand by and watch as they are lost. As Dr. Terry Shintani has told me of what he has gained through his years of working in the Hawaiian community of Waianae and encountering similar situations elsewhere in the world, "The health problems of Native Hawaiians are reflective of what happens to all people when they abandon the diet and ways of their ancestors."

Fully facing the implications of these problems forces us to make choices about our own lives. Each of us must ultimately make these tough choices for ourselves, and on behalf of our loved ones. Competing options are sometimes placed before us, and in periods of anxiety or grief, we are not always best able to understand the ultimate consequences of the option we choose.

So it is today when doctors diagnose us with a disease that they say may be in our genes, but that also affects and is affected by our nutritional regimen. This can happen to any of us, not just Native Hawaiians, but Ashkenazi Jews, Mennonites, Zunis, Senegalese, or Mongolians as well. The smaller the size of our remaining ethnic population, and the longer we have intermarried among ourselves, the higher the probability that some doctor somewhere is consulting with our newly married couples about how to avoid "birth defects." There may be a moment of relief when someone has determined just what has been ailing us, but there is often a second wave of anxiety associated with deciding just what to do with the diagnosis. A doctor or team of doctors and geneticists can recommend that we take an expensive medicine day after day for the rest of our lives, or that we shift our diet and exercise regimes, or that we consider some kind of gene therapy.

The number of gene therapies available to us has multiplied exponentially over the last few years, thanks to many recent breakthroughs in genetic research. By any measure, the mass of genetic information about the human condition developed since James Watson and Francis Crick elaborated the structure of DNA in 1958 is astonishing. By the year 2000, well over 97 percent of the human genome had been mapped, and more than 2.5 billion base pairs of DNA had been sequenced. The implications for describing and diagnosing what doctors refer to as genetic "disorders" are only now being fully realized.

The richness of information suddenly available through genetic screening offers diagnoses of such fine detail that some patients are overwhelmed with hope—hope that the solution to their problem will be as straightforward as their screening was. Via your doctor, a geneticist could tell you something like this:

"You carry a rare allele on gene p on chromosome 2 that results in the lack of production of a certain enzyme needed to metabolize fat efficiently. The best current evidence is that this deficiency is more frequently found among people with ancestors who left central Asia for northeastern Africa some 5,000 to 10,000 years ago. It is now found in one out of every four people of your ethnic background, typically among homozygous males. While not lethal, this enzyme deficiency makes you prone to accumulating fat in your body and plaque in your arteries. In other words, it is an independent risk factor for heart disease, the number one killer of people your age. Without any intervention, it is estimated that half of all homozygous carriers of this allele may suffer from a shortened life span."

Your mouth goes dry, and you glance at your family, noticing that tears are welling up in their eyes. You squeeze your wife's hand, and ask the doctor what your options are.

"Well, we can give you a rather expensive cholesterol-lowering drug that I'm afraid you'll have to take for the rest of your years here on earth," he states coolly, but then he tries to make light of the life sentence. "Of course, that's why the two of you pledged to stay together in sickness and in health, at the pharmacy or in the weight loss program, till death do you part."

"So she's stuck with an overweight husband who needs regular assistance in administering his medication," you say. "Any other options besides divorce? That might at least reduce the frequency of my genes in the population."

"Well, I could put you on a stringent diet and exercise regime, but it takes a lot of self-discipline. So instead, I suggest that you consider a new gene therapy."

The doctor then describes a new procedure by which a virus carries a gene into your body that allows your production of that scarce enzyme to increase exponentially. He calls it a "gene repair" strategy and likens it to emergency mechanics diving down to help a submarine fix its pressure chambers so that its crew can ascend once more for air.

"A virus? You're gonna inject a virus into my body?"

"Not now. And don't worry, it's not like a cold or flu virus. Anyway, it's still several years away from commercial release, and it's likely to be expensive, so save your pennies, take some meds, and get a lot of exercise until then. Hopefully, it will soon provide people like you a third option to pharmaceutical and nutritional therapies."

This emerging set of biotechnologies—collectively known as *functional genomics* or *nutritional genomics*—will surely help some people in the future, but such therapies are also likely to inadvertently generate both health problems and ethical dilemmas for many others. In the short run, functional genomics will not necessarily become accessible to all, because subtle remnants of racism that persist around us may simply mean that doctors do not offer therapy to all who may benefit from it, or may not make all their patients comfortable when describing its risks and potential benefits. Some who might potentially benefit may never even hear of such advances, especially those living outside of the industrialized Western countries, or people within those countries, who live in some of the poorest areas, such as sharecroppers in the Deep South; the people who may most desperately need the promise of functional genomics might live in areas remote from cutting-edge medical research. Language barriers and more pressing health concerns may prohibit others from partaking of the latest medical knowledge; you may have distant, monolingual kin who dwell along the Nile, where schistosomiasis and AIDs are more threatening than slowly developing cardiovascular disease. Even if genetic screening were accessible to these people within a decade or two, would they be able to afford the cost of travel and screenings to gain access to gene therapy?

Compared to the "digital divide," which provides the haves with high-speed Internet access and the have-nots with no computer access at all, the "genomics divide" will be even more difficult to bridge.

Even if genetic screening were to suddenly become available to all people—as sickle-cell anemia screening did for most African Americans in the early 1970s—it is not a given that such information will be used in a morally and ethically sound manner. Consider what happened with this sickle-cell screening just a quarter century ago, and it becomes clear that screening is not necessarily a politically, socially, or economically neutral activity, nor are its consequences always benign.

Dr. Robert Murray Jr., who has headed up Howard University's Medical Genetics Division and has served as a member of the U.S.

National Academy of Sciences, recalls that not too long ago, both government and private corporations were abusing the availability of genetic information about sickle-cell carriers:

> In many cases, the information resulting from [screening] programs was not used to benefit the people who were tested but was used in ways that harmed them. People were unnecessarily excluded from high-risk but desirable positions. Military personnel with the trait were excluded from service as paratroopers, SCUBA divers, underwater demolition experts, and [from] submarines. . . . Problems occurred in the civilian sector as well. For example, after a report suggested that life expectancy for people with the trait was 5 percent shorter than for those without the trait, insurance companies raised their premiums for people who were carriers of sickle-cell genes. A survey found that twenty-seven companies took this completely unfair and unjustified action. Publicity about sickle-cell anemia also led to employment discrimination (Murray 2001).

Nutritional anthropologist Fatimah Linda Collier Jackson has also noted that such discrimination generally ignores that genetic information is typically not sufficient in and of itself to predict disease risk or life expectancy:

> The relationship between genetics and disease is not a linear one. It is rarely the case of, 'If you have the gene, you have the disease.' Rather, there is nuance. There are gene-environment interactions. There are gene-gene interactions. All these components need to be taken into ac-

> count before we can set down some hard and fast rules about who is the normal [i.e., healthy] human (Jackson 2001).

Fortunately, there are people on the far side of the genomics divide who are not waiting for epidemiologists to decide if genes will be the deciding factor in "normalcy" or "health." Instead, they are taking their fate—via their traditional foods—back into their own hands. Rather than waiting for some supposed silver bullet like gene therapy to come along and save them, they are improvising real-life solutions that take genes into account, while also drawing upon ancient cultural food traditions and inspiring community support networks as well. Some of these communities are easily among the poorest of the poor in America, Australia, and on other continents and islands. But this has not stopped them from finding a way to *eat in place* that complements rather than conflicts with their genes. To understand why genetically influenced nutritional maladies are not the inevitable fate of all who may be predisposed to them, one need only go to Hawaii. There we can see how homegrown solutions are fully compatible with emerging theories regarding dietary diversity that are being implemented to restore the health of ethnic populations.

~ Until recently, if you visited the Waianae Coast of Oahu and looked for something wholesome to eat, you would find few satisfying choices. Driving along the Farrington Highway edging the coast, you would pass the fastest and fattest of the food chains: Kentucky Fried Chicken, Burger King, McDonald's, Long John's Pizza, and Pizza Hut.

There are a few nonfranchised minimarts and drive-ins—owned mostly by Filipinos—which feature barbecued chicken backs and thighs cooked on a huli-huli spit, smoked over mesquite wood. Alongside them, you might be able to buy *pasteles*, *gandules*, or the local analog of Krispy Creme donuts, the deep-fried *malassadas*. At first glance, it would seem that residents of the Waianae Coast do not live off the fat of the land, they have imported the fattest of the fat from other lands, for the coast is now one of the great melting pots of the world, with Polynesians, Asians, Africans, Europeans, and Latinos all intermixing. Until recently, that fat had slowly, inexorably built up in residents' bodies, arteries, and veins. And as exotic fatty foods spread along the coast, a variety of maladies caused by malnourishment got a foothold in paradise.

The friend who introduced me to the Waianae Coast, Dr. Terry Shintani, has summed it all up, saying, "The tragic irony is this: While Hawaii is statistically the healthiest state in the U.S., the Native Hawaiians have had the worst health in the nation . . . 2½ times the heart disease, 2 times the cancer, 2½ times the strokes, 7 times the diabetes, and 4 times the infectious disease mortalities compared to all races in the U.S."

By the year 2000, one in five Native Hawaiians suffered from diabetes, and half were labeled obese by their physicians. Worse yet, diabetes-related mortalities have been six times higher among Native Hawaiians than among the U.S. population in general.

The poster child for this set of problems among Native Hawaiians has been the beloved Hawaiian musician Israel "Iz" Kamakawiwo'ole, a man with enormous talent, a huge body, and an even greater vulnerability to diabetes and heart disease. By the time he was in his thirties, Iz was a celebrity not only among his own people, but in the world music scene abroad. One can only wish that Iz had found enough nutritional help early on before he topped the scales at 757 pounds. His artistic contributions were cut short at age 38, as diabetes and other maladies brought him down. When they lowered his massive body by crane into a viewing area at Hawaii's state capitol, more than 40,000 Hawaiians came to pay their respects, grieving the loss of a brother who could sing like a bird but had never learned to make appropriate food choices for his genotype.

Fortunately, the Waianae Coast has recently been blessed with some "new" options, ones that offer hope and health where there might otherwise be despair. One of these is the Waianae Coast Comprehensive Health Center's new dining pavilion on its Native Hawaiian Healing Center Grounds. The other—smack dab in the middle of the town of Waianae—is a small but lovely cafe recently renovated and reopened by Mala 'Ai 'Opio, a nonprofit with a five-acre farm nearby.

But to understand how these options emerged as hard-won victories for Hawaii's largest native community—after decades of tragedy—it is worth retracing the tracks leading to the Native Hawaiian Renaissance Movement of the 1970s.

During that tumultuous era, the Waianae Coast was perhaps even more economically impoverished than it is today. The sugar mills had closed, leaving employment at nearby military bases one of the few viable means of earning wages in the area. And yet, those bases also generated a certain bitterness among the locals, for they had tapped and diverted springs high in the mountains above the coast, keeping water from flowing into Native Hawaiian taro fields as it had done for centuries. Vietnam War protests and disappointment with the failed promise of prosperity following Hawaiian statehood also contributed to the local unrest.

Worse yet, the Native Hawaiian population was close to being swamped. Compared to its peak size of 400,000 to 800,000 on all the islands just prior to the arrival of Captain Cook in 1782, there were perhaps only 2,000 pure-blood Native Hawaiians surviving into the 1970s. Many of these *piha kanaki maoli* felt as though the familiar world around them was slipping from their hands. Even the larger mixed-blood native population—the *hapa kanaka maoli*—had been beaten back over the seven decades in which their language had been banned in schools and newspapers, the practice of their medicine outlawed, and their tenure to traditional *ahupua'a* foodshed management units gradually taken from them, from mountain ridge to sea shore.

Land loss, language loss, culture loss, and weight gain. These trends seemed unstoppable to many until Eric Enos and others in the Native Hawaiian Renaissance uncorked the creative energy of their own people, disconnected the pipes that had once carried the water away from them, and tapped into the healing powers of the plants all around them. By 1979, Enos and his friends had given birth to the Cultural Learning Center at Ka'ala Farms, nestled in the foothills of

Oahu's mother volcano just above the Waianae Coast. The Center revived traditional Hawaiian values and cultural practices, including the cultivation of *lo'i kalo*, or paddies of the ancient root crop, taro.

Solomon Enos, Eric's son, who is now a leader in the second wave of this renaissance, once tried to explain to me what it was like growing up during that period of cultural restoration and experimentation. He smoothed back his long black hair, and laughed, as we talked among the *kalo* plants now growing in Waianae.

"With the help of some University of Hawaii activists who brought in pipes and other supplies, my father undid the government's diversion of this big spring in the mountains and redirected it back to the overgrown *lo'i kalo* below. All these kids were there to help him clear the land and reopen the paddies, right there in what was once the 'poi bowl' of Oahu, in the mother caldera."

By "poi bowl," Solomon referred to some two hundred acres that once grew the taro roots pounded into poi, the purplish, nutritious mush that was a mainstay of all Polynesians, not just the Hawaiians of Oahu.

Where Solomon and I spoke was not twenty miles away from Ka'ala Farms, where we weeded vegetable and root crops at another farm engendered by the Native Hawaiian Renaissance. As he showed me the varieties of taro that they grew there, he told me that it was with that root crop that his own roots lie.

"So I grew up in and around taro paddies, *lo'i kalo*, as my father ran this camp for troubled youth where we cleared and reclaimed ancient fields. Our own word for the baby taro or *keikei* offshoot that we pull off the mother corm of the taro is *'ohā*. That's where we get our word for family or kin, *'ohana*."

I imagine that the taro paddies were a great place to begin to restore the roots of Native Hawaiian communities. In precolonial eras, some 50,000 to 60,000 acres of hundreds of taro varieties provided the staple crop for well over a quarter million Hawaiians. Worldwide, perhaps 100 million people still rely on taro and other similar root crops as their nutritional mainstay. By 1900, however, taro paddies had declined to less than 20,000 acres across all the Hawaiian Islands, and by 1980, only 500 acres were left in regular cultivation on the islands. As cultural ecologist Nan Greer has documented, declines in taro paddies appear to have contributed to declines in endemic waterbirds that formerly frequented the *lo'i kalo* habitats; but as these traditionally managed wetlands are restored, the now-endangered birds should recover in number. In other words, Greer hypothesizes, the restoration of *lo'i kalo* is as good for the wildlife as it is for the people of the land.

When Eric and his fellow activists began to revive taro, they had a gut sense of how delicious and nutritious it was, but recent research has borne out their intuitions. As one of but a handful of staple crops that are hypoallergenic, taro is rich in calcium, potassium, iron, phosphorus, thiamin, riboflavin, several other B vitamins, as well as vitamins A and C. It contains no fat or cholesterol, but does contain enough soluble fiber and amylase starches to function as a slow-release food, just as many traditional foods from the Australian and American deserts do. And so, Eric and friends had chosen an appropriate keystone with which to rebuild the arch of Native Hawaiian culture.

That is where the Waianae Coast Comprehensive Health Center comes in, founded just three years after Ka'ala Farms in 1982. It soon became the largest health-care provider for Native Hawaiians. As its current director, Richard Bettini, recalls, "The founders had always wanted to bring the best of allopathic medicine to this low-income community, but about 13 years ago, we decided that we must also integrate traditional Hawaiian cultural values, beliefs, and practices into what we do. We realized that there was something to Native Hawaiian culture that might otherwise be lost, and we wanted to bring it back into our approach to health."

By 1987, the Center's staff was hearing about the success that the Na Puuwai group on another island had been having in reducing cholesterol and heart disease risks with the Molokai Diet of traditional Hawaiian foods. The Center brought in from Molokai a remarkable woman, Helen Kanawaliwali O'Connor, who had worked with Na Puuwai on that project and was willing to help them as well. O'Connor had come from a family of traditional healers of *kanaki maoli* ancestry and had never needed to go to a hospital or Western-trained doctor her entire childhood. As her friends on the Waianae Coast now appreciate, O'Connor was key to the integration of Native Hawaiian healing with Western medicine for a very basic reason: she had a great talent for listening to others, and listening deeply.

"Helen is gifted at what we call 'talking story' with patients," said Richard Bettini, "so that they are assured someone here at the Center truly hears what they feel their problems are."

Arriving at the Center around that same time was one of the most brilliant and compassionate health-care practitioners I have ever encountered anywhere, Dr. Terry Shintani. Of Japanese ancestry, but now fully adopted into the Kanahele family of the Waianae Coast, Shintani has medical and law degrees from the University of Hawaii, as well as a masters in nutrition from Harvard. But what he is best known for is advancing the (re)integration of diet with genes and culture in health-care practice, rather than treating them as separate considerations. With O'Connor's help, he began to rough out a diet based on traditional Hawaiian foods that not only reduced cholesterol and the risk of heart disease, but dealt with diabetes as well.

Shintani is of small stature compared to most Hawaiians, but he nevertheless looms large in any room, for his winning smile, quick wit, and broad interests seem to fill up any vacant space. He has spent years studying the various orally transmitted healing traditions of the *kanaki maoli* with his adopted brother, Kamaki Kanahele, and their mother, Auntie Aggie, both healers in their own right.

"This diet isn't something I learned at Harvard," he told me one time. "It's something that Hawaiians and their ancestors knew for thousands of years. They knew that food without *mana*—that is, without *life force*—is not going to support anyone's health."

He paused for a moment and looked me in the eye, making sure I understood that he was not merely romanticizing Hawaiian traditions. Instead, he was attempting to honor them for their underlying principles, principles that distilled thousands of years of experience regarding why people get sick.

"In traditional medicine," he said, "it is recognized that there is really only one disease that all of us must learn to resist: *arrogance*. It is simply arrogant to think that we can violate the laws of nature and get away with it."

And so, around 1989, O'Connor, Kanahele, and others began to help Shintani shape an approach to diet based not merely on taro but on a Hawaiian understanding of the laws of nature—a diet that could potentially restore their community's health. They decided, in Shintani's words, that "if the problem was a nutritional one, the answer wasn't more medication, it was diet. When [Native Hawaiians] regularly ate the old foods, they didn't have these diseases. So we went to the community to learn more about taro and the other old foods."

Many of the foods considered by Native Hawaiians to have health benefits were rich in what Shintani now calls the "good" carbohydrates—the slow-release ones. But his documentation of their value in controlling obesity and diabetes ran counter to what Dr. Atkins and many other popular diet gurus claimed was the best diet for *all* overweight individuals at risk for diabetes—a diet *low* in plant carbohydrates and *high* in protein and fats from free-ranging livestock or wild game and fish. Seldom one to worry about running counter to prevailing fads, Shintani had begun to see that diets high in good carbohydrates and low in animal fats were *working* among his Native Hawaiian patients.

"Our Hawaii Diet™ is based on traditional diets," he told me, "and recognizes that individuals may be adapted to various dietary patterns. [It is based] on the principle that many diet-related diseases could be reversed by returning people to their traditional culture-based diets."

Shintani realized that the principles he and his colleagues were following had applications among other ethnicities as well, but his team has focused on reaping the benefits of traditional Hawaiian foods for Native Hawaiians. While acknowledging that rapid diet change has universally gotten people out of sync with their genes, Shintani's early writings also hinted at the special circumstances behind the special nutritional needs of Native Hawaiians—circumstances rooted in the nature of island biogeography: "Adaptation of an island population to a particular set of foods in a diet (combined with lifestyle) may be just as important as food composition in health . . . [for] there are genetic differences among people's body responses to food via blood sugar, cholesterol, allergies, weight gain, etc." (Shintani et al. 2001).

While Shintani was distilling such reasoning into a taro-based diet that could be tested for its efficacy in controlling diabetes, his colleague, Sheila Beckman, was developing the research protocols that would allow them to compare the precursor to the Hawaiian Diet™, the Waianae Diet, against others. She settled on a regime that offered 1,569 calories a day, with 78 percent of its energy as carbohydrates, 15 percent as protein, and 7 percent as fat. It included not only taro, but sweet potato, fern shoots, seaweed, native fruits, fish, and fowl. Beckman developed an experimental design that kept twenty people fed for twenty-one days using a flexible all-you-can-eat menu of traditional Hawaiian foods.

"Exercise, though we knew that it was also as important," Beckman recalled to me, "was not formally integrated into the 1991 study, only diet. We were including some fairly large Native Hawaiians from the homestead lands up the valley, who wanted to be included but initially had limited capacity for a lot of physical activity. Of course, they were the ones that lost the most weight, I guess because they had so much they could lose. Later, we incorporated exercise in the follow-ups with them."

Among those first participants in evaluating the benefits of the traditional Hawaiian diet were a number of prominent community leaders. While concurring with Beckman that such a test required strict research protocols, Shintani's mind was already racing ahead to the *cultural inspiration* that their modest experiment might generate.

"I incorporated one key insight I had gained from my days as an activist and organizer in the seventies," he told me, "work with the leaders of the community, who others look up to. Even in that first study, we invited Native Hawaiian leaders from the homestead lands in the four neighboring valleys to be part of the diet's implementation. We met with them for morning prayers and songs before breakfast. Then all of us ate together. For lunch and for any between-meal snacks, they could take with them all the Hawaiian foods they wanted. We gathered back together for a meeting in the evening, eating as a group once again, talking through how everyone felt."

Helen Kanawaliwali O'Connor helped facilitate these "talking story" sessions, on top of staying up many nights stirring the taro, making poi. O'Connor enabled her neighbors to comfortably talk through all that they had been experiencing, mixing in anecdotes from their childhood, admonitions from their elders, and traditional Hawaiian wisdom. She *listened*, rather than cutting them off. She was a staffer and yet a participant as well, for she, too, experienced what the others were witnessing: rapid weight loss and a dramatic drop in blood sugar that reduced their need for insulin by eighty units in five days.

Shintani told to me how the community responded to the stories that their leaders brought home: "It was like a lightning bolt ran through the community, revealing something that had always been out there in the darkness that most of them hadn't seen before. That excitement alone has generated so many lasting effects."

When the Waianae Diet results were finally published in 1991 in the *American Journal of Clinical Nutrition*, the local communities were not the only ones standing up and taking notice. The scientific community was just as stunned by the good news: participants lost an average of 17 pounds in 21 days; their cholesterol levels dropped 12 percent; and their blood-sugar levels dropped by 26 percent, bringing most of the participants into the "safe zone" for blood sugar–insulin interactions, making regular use of hypoglycemic pills or other interventions unnecessary.

Perhaps even more heartening were the results that Beckman and Shintani accumulated over the following eight years from some eighty-two participants in various trials of the Waianae Diet. Dieters maintained an average weight loss of 15.1 pounds over the seven and a half years of periodic monitoring. One exceptional individual shed 174 pounds. Another dropped "only" 117 pounds, but *kept it off over the following eight years*. Although the average duration of surveillance was thirty-four months, two-thirds of all participants continued to weigh less than when they began the Waianae Diet in earnest.

These are phenomenal successes for any kind of long-term weight intervention effort, or for that matter, any community-based initiative to change local patterns of consumption.

For me, the most revealing result of the Waianae Diet—at least in terms of gene-diet interactions—has never been published in a way that explicitly discusses its significance. A decade after running the first cohort of Native Hawaiian participants through the diet, Shintani, Beckman, and O'Connor took a multiethnic group of twenty-two individuals through a comparable study. Because participants were from a variety of cultural backgrounds, they were allowed to eat foods from their own ethnic traditions that more or less fit the same criteria as the traditional Hawaiian foods. In other words, the total number of calories and the percentage of energy from carbohydrates, protein, and fat were held around the same levels, but the particular foods eaten varied with one's own cultural and individualistic taste preferences.

This broadened adaptation of the original Waianae Diet's high-carbohydrate/low-fat regime also resulted in dramatic weight loss, as well as improved blood pressure, blood-sugar, cholesterol, and low-density lipid levels. Shintani was certainly justified in suggesting that such a regime might benefit far more people than Native Hawaiians alone. Because the 1991 and 2001 studies followed essentially the same protocols, ran for the same duration, and had roughly the same number of participants, I decided to quantitatively compare the results of the 2001 multicultural cohort with data from the original 1991 Native Hawaiian group.

While the Native Hawaiian cohort lost an average of 17 pounds, the multicultural cohort lost "only" 10.8 pounds, even though the latter group had formally incorporated exercise into their schedule from the very start of their twenty-one day effort, something the Native Hawaiian cohort had not necessarily done. Nevertheless, triglyceride and blood-sugar levels had dropped far more dramatically among the Native Hawaiians. Most other indicators of success were comparable for the two groups, with the exception that blood-pressure levels among the multicultural cohort dropped more precipitously. As in Jennie Brand-Miller's comparison of Australian Aborigines and Australian Caucasians (see chapter 7), the health benefits of slow-release foods were significant for both groups, but the greatest improvements were observed in the indigenous peoples returning to a traditional diet that had perhaps been abandoned by their ancestors only a couple of generations ago.

It might be worth remembering the motivation for the Waianae Coast's community-based efforts to combat the entire Syndrome X cluster of health-risk symptoms. Shintani and his colleagues have been spurred by a deep-seated conviction that the issue in their community requiring resolution is larger than a "genetic disorder." Just as local residents desperately wanted their own bodies to be healthy once more, they also wanted healthy spirits, a healthy community, and healthy land surrounding them. None of these goals outside of the physical one can be achieved merely through medication or gene therapy. In fact, some residents of the Waianae Coast would argue that their bodies could not be maintained in good health unless their culture and their habitats are brought back to full health as well. What good is knowing that taro, sweet potato, or other traditional crops can heal your body if no one around you is interested in growing them anymore in a way that can build community and build fertility in the Waianae landscape? I was at last beginning to see the whole, not just the isolated parts of the interactions among genes, particular foods, and specific cultural traditions.

And so, while visiting the Waianae Coast, I spent only part of my time at the health center; the rest was spent with Solomon Enos, Kukui Maunakea-Forth, and Gary Forth at the Mala 'Ai 'Opio farm several miles up the road. Like Ka'ala Farms, Mala 'Ai 'Opio provides a safe harbor for youth still finding their way into the adult world, but instead of growing taro in paddies, Mala 'Ai 'Opio incorporates mixed-crop fields with two dozen other root, fruit, and leafy vegetable crops. These crops are destined for the group's new restaurant in Waianae, but will likely be used at the health center as well. Organically grown by a group of youth that learn cultural ethics, songs, and prayers while they work, Mala 'Ai 'Opio's produce is about healing at all levels and resisting the forces of globalization that might otherwise diminish local food traditions.

After working a morning planting seeds and weeding, praying, and laughing with the Mala 'Ai 'Opio farm crew, I had a chance to experience how the many strands of the Waianae Coast come together. As Kukui Maunakea-Forth told me as we sat down together for the inauguration of the new healing grounds at the health center, "a lot of us see one another pretty frequently, because we're on each other's citizen's advisory boards, whether it's a health project, a farming project, or a cultural education project." Perhaps that is the peculiar strength of the Waianae Coast's indigenous community, which appears to be an essential element of any successful community health program: its farmers, foragers, fishermen, educators, traditional healers, physicians, chefs, nutritionists, and activists are all rowing in the same direction. In this sense, the Waianae Coast community does not use the term "comprehensive health," in an idle or superficial manner. The health of the lands and waters, of the culture and the community, is not separate from the health of individuals.

Those of us attending the ceremony had climbed up a volcanic ridge overlooking the sea, one tier above the little clinic where the Waianae Coast Comprehensive Health Center began thirty years before; the Center was celebrating its anniversary by dedicating the new healing grounds, with its gardens and dining pavilion. As we arrived, the clinic's staff placed leis of fresh flowers around our necks and kissed us. We took our places along stone benches in a garden of healing food and medicinal plants overlooking the ocean. When the benches and chairs were packed tight with community members, the deep bass sound of a conch shell blew us back into a more ancient time, or perhaps, into a *timelessness*. The traditional conch blowers stood on the balcony of the newly constructed dining pavilion high above. Then we heard the sound of chanters moving toward us—Kamaki Kanahele's towering figure, with his long braided ponytail facing us as he gestured and sang an invocation to the other chanters. They responded to his invocation, streaming down the ridge until they filled the remaining space in the gardens.

These singers of traditional Hawaiian chants—the nurses, doctors, van drivers, groundskeepers, X-ray technicians, and counselors of the clinic—had been practicing for months under Kanahele's direction to sing for this inauguration:

> *E ko makou mau kia'I msai kalani mai,*
> (O our ancestors from remote antiquity,)
>
> *E nana ia mai ka hale ame ka aina,*
> (Watch over our house and land . . . )
>
> *Mai ka uka a he hai*
> (From mountain to sea
> From inside to outside)
>
> *Kia'I'a, malama ia*
> (Watch over and protect it)
>
> *E pale aku I na ho'opilikia ana, I ko kakou nohona*
> (Ward off all that may trouble our life here)
>
> *Aloha e, aloha kakou e, aloha e*
> (Aloha!)

Following chanting and hula, orations and blessings, we slowly strolled up to the dining pavilion. There with Shintani, Kanahele, Kukui Maunakea-Forth and others, we filled our plates with slices of steamed taro corms and sweet potatoes, bowls of poi, and strips of meat or fish wrapped in taro leaves. Taking our plates out to the balcony, we watched the giant waves stir the seaweeds and redirect the schools of fish out in the distance. When I turned around and looked back into the dining pavilion, I saw a sea of jubilant Native Hawaiian faces. They were pleased—if not jubilant—to once again be *eating in place, eating with their ancestors, and eating what was fit for their genes and their cultural identity*.

∼ It is significant that the greatest health improvement I have witnessed in any community has come from one in which gene-food interactions have been positively influenced without a reductionist focus on either the genes or the diet. Instead, the Waianae community has built a larger set of positive relationships within which gene-food interactions are nested. The community is giving its members incentives to improve their health through a variety of mutually reinforcing means.

Of course, it may be tempting for epidemiologists as well as nutritionists to dismiss what the Hawaiians have achieved as being merely a feel-good story. Cynics might claim that while the Hawaiians have restored their traditions, this effort has not necessarily been informed by the best cutting-edge health science, nor by the haunting statistics on the decline in health of various groups worldwide. But they would be wrong.

Hawaiians listen just as deeply to those two views of the world as they do their own traditions. The scientific inquiry into the value of native Hawaiian foods is indeed cutting edge, and Waianae community members are actively engaged in exchanges with virtually every major indigenous culture that is dealing with similar health issues. I can only hope that the rest of us will learn to listen as fully to what the Hawaiians, and many other ethnic communities, are telling us: *be aware of the risks and grieve the losses; ethically use both traditional knowledge and the best available science you can find in a manner that honors the contributions of both; but at the same time, renew the vital connections between your body and the land that are essential to restoring health at all levels*.

Modern science will no doubt continue to enrich us with many new insights about the connections between genes, diets, and disease. Parts of this story will need to be revised, as new findings alter the picture as a whole. But science alone cannot ensure that we will grow healthier simply because health professionals learn more about gene-food interactions. Each of us must also take the time to turn inward, to reflect upon our family histories of eating, exercising, and evading disease. We must gain a deeper sense of what has tended to make us sick and what has served to keep us well—connected to our community, culture, and homelands. And we must act to protect those connections between food, heritage, and habitat that underlie every moment we live healthfully and happily on this diverse planet.

# SOURCES

I wish to acknowledge the following people for their inspiring discussions of gene-food interactions: Laurie Monti, Carlos Martinez del Río, Josh Tewksbury, Janette (Jennie) Brand-Miller, Loren Cordain, Nina Etkin, Sol Katz, Terry Shintani, Paul Rozin, Tim Johns, Antonis Kafatos, Amadeo Rea, Aglia Kremezi, Alfredo López Blanco, Mike Mitchell, Ruth Giff, Sally Pablo, Nikos and Maria Psilakis, Antonia Trichopolou, Arno Motulsky, Bill McKibben, Rob Robichaux, Fatimah Linda Collier Jackson, Helen Kanawaliwali O'Connor, Linda Bartoshuk, Solomon Enos, Kamaki Kanahele, Barry Infuso, Nan Greer, Boyd Swinburn, Paul Sherman, Jim Berry, Charles Weber, Steve Colaguiri, Paul Bosland, Eric Votava, and Suzanne Morse. Further thanks to many members of the communities of Denpasar and Ubad, Bali; Spili, Crete; Oristano, Sardinia; Athens, Greece; Ak-Chiñ, Sells, and Sacaton, Arizona; Punta Chueca and Desemboque, Sonora; and Waianae, Hawaii.

Sarah Jane Freymann, Barbara Dean, and Naima Taylor were of inestimable value in keeping this project moving along. As she has in the past, Agnese Haury also provided support in many ways. This project also benefited from a U.S. National Science Foundation/Australian Board of Science grant to me and Janette (Jennie) Brand-Miller to discuss native diets and diabetes at Kims Toowoon Bay, New South Wales.

## INTRODUCTION

Barbujani, G., and L. Excoffier. 1999. The history and geography of human genetic diversity. In *Evolution in health and diseases*, ed. S. C. Stearns, 27–40. Oxford: Oxford University Press.

Bland, J. S., and S. H. Benum. 1999. *Genetic nutritioneering.* Los Angeles: Keats Publishing.

Grierson, B. 2003. What your genes want you to eat. *New York Times Magazine*, May 4, 77–79.

Harmon, D. 2002. *In light of our differences*. Washington, DC: Smithsonian Institution Press.

McKibben, B. 2003. *Enough: Staying human in an engineered age*. New York: Times Books.

McKusick, V. A. Accessed 2003. Online Mendelian Inheritance in Man database. www.ncbi.nlm.nih.gov/entrez/query.fcgi?db=OMIM.

Moore, D. S. 2001. *The dependent gene: The fallacy of "nature" vs. "nurture."* New York: Times Books / Henry Holt.

## CHAPTER ONE

Agarwal, D. P., and H. W. Goedde. 1986. Ethanol oxidation: ethnic variations in metabolism and response. In *Ethnic differences in reactions to drugs and xenobiotics*, ed. W. Kalow, H. W. Goedde, and D. P. Agarwal, 99–112. New York: Alan R. Liss, Inc.

Brown, C. 2000. *The ghosts of evolution*. New York: Basic Books.

Grierson, B. 2003. What your genes want you to eat. *New York Times Magazine*, May 4, 77–79.

Jackson, F. L. C. 1991. Secondary compounds in plants (allelochemicals) as promoters of human biological variability. *Annual Review of Anthropology* 20:505–546.

Katz, S. H. 1990. An evolutionary theory of cuisine. *Human Nature* 1 (3): 233–259.

Kretchmer, N. 1972. Lactose and lactase. *Scientific American* 227:70–78.

Lieberman, M., and D. Lieberman. 1978. Lactase deficiency: A genetic mechanism which regulates the time of weaning. *American Naturalist* 112:625–639.

Long, J. C., W. C. Knowler, R. L. Hanson, M. Urbanek, E. Moore, P. H. Bennett, and D. Goldman. 1998. Evidence for genetic linkage to alcohol dependence on chromosomes 4 and 11 from an autosome-wide scan in an American Indian population. *American Journal of Medical Genetics* 81 (3): 216–221.

Ridley, M. 2000. *Genome: The autobiography of a species in twenty-three chapters*. New York: Harper-Perennial.

Rozin, E. 1989. The structure of cuisine. In *The psychophysiology of food selection*, ed. R. Shepard, 189–208. New York: John Wiley and Sons.

Rozin, P. 1982. Human food selection: The interaction of biology, culture and individual experience. In *The psychobiology of human food selection*, ed. L. M. Barker, 225–269. Westport, CT: AVI Publishing.

Saavedra, J. M., and J. A. Perlman. 1989. Current concepts in lactose malabsorption and intolerance. *Annual Review of Nutrition* 9:475–502.

Sherman, P. W., and J. Billings. 1999. Darwinian gastronomy: Why we use spices. *BioScience* 49 (6): 453–463.

Simoons, F. J. 1973. The determinants of dairying and milk use in the Old World: Ecological, physiological and cultural. *Ecology of Food and Nutrition* 2:83–90.

Wade, N. 2002. As scientists pinpoint the genetic reason for lactose intolerance, unknowns remain. *New York Times*. January 14, 2002.

Williams, G. C., and R. M. Nesse. 1991. The dawn of Darwinian medicine. *The Quarterly Review of Biology* 66 (1): 1–81.

## CHAPTER TWO

Cordain, L. 2002. *The Paleo Diet*. New York: John Wiley and Sons.

Cordain, L., J. C. Brand-Miller, S. B. Eaton, N. Mann, S. H. A. Holt, and J. D. Speth. 2000. Plant-animal subsistence ratios and macronutruient energy estimations in worldwide hunter-gatherer diets. *American Journal of Clinical Nutrition* 71:682–692.

D'Adamo, P. J., and C. Whitney. 2002. *Eat right for your type*. New York: Riverside Books.

Darwin, C. 1859. *The origin of species by means of natural selection*. Chicago: Encyclopedia Brittanica (reprint 1990).

Eaton, S. B., and M. Konner. 1985. Paleolithic nutrition: A consideration of its nature and current implications. *New England Journal of Medicine* 312:283–289.

Eaton, S. B., M. Shostak, and M. Konner. 1988. *The Paleolithic prescription*. New York: Harper and Row.

Eaton, S. B., S. B. Eaton III, M. J. Konner, and M. Shostak. 1996. An evolutionary perspective enhances understanding of human nutritional requirements. *Journal of Nutrition* 126:1732–1740.

Ehrlich, P. R. 2000. *Human natures: Genes, cultures, and the human prospect*. Washington, DC: Island Press.

Grierson, B. 2003. What your genes want you to eat. *New York Times Magazine*, May 4, 77–79.

Jackson, F. L. C. 1991. Secondary compounds in plants (allelochemicals) as promoters of human biological variability. *Annual Review of Anthropology* 20:505–546.

Lewontin, R. C. 1998. *Human diversity.* New York: Scientific American Library.

———. 2004. *The triple helix: Gene, organism and environment.* Cambridge: Harvard University Press.

Morewood, M. 1997. Quoted in P. Van Oosterzee, *Where worlds collide: The Wallace line.* Ithaca: Cornell University Press.

Olson, S. 2002. *Mapping human history: Discovering the past through our genes*. Boston: Houghton-Mifflin.

Reaven, G., T. K. Strom, and B. Fox. 2001. *Syndrome X: The silent killer*. New York: Simon and Schuster.

Satel, S. 2002. I am a racially profiling doctor. *New York Times Magazine*, April 13, 56–60.

Sears, B. *The Zone Diet*. New York: Harper-Perennial.

Simopoulos, A. P., and J. Robinson. 1999. *The Omega Diet*. New York: Harper-Perennial.

Somer, E. 2002. *The Origins Diet*. New York: Owl Books / Henry Holt.

Strassman, B. I., and R. I. M. Duarte. 1999. Human evolution and disease:

Putting the Stone Age in perspective. In *Evolution in health and disease*, ed. S. C. Sterns, 91–101. New York: Oxford University Press.

Wallace, A. Quoted in P. Van Oosterzee, *Where worlds collide: The Wallace line*. Ithaca: Cornell University Press.

Weiner, J. 1994. *The beak of the finch: A story of evolution in our time*. New York: Alfred Knopf.

Williams, G. C., and R. M. Nesse. 1991. The dawn of Darwinian medicine. *The Quarterly Review of Biology* 66 (1): 1–81.

## CHAPTER THREE

Andrews, A. C. 1949. The bean and Indo-European totemism. *American Anthropology* 51:274–291.

Brown, P. J. 1979. Cultural adaptations to endemic malaria and the socio-economic effects of malarial eradication in Sardinia. PhD. dissertation in Anthropology, State University of New York at Stony Brook.

———. 1986. Cultural and genetic adaptations to malaria: Problems of comparison. *Human Ecology* 14 (3): 311–329.

Carson, P. E., C. L. Flanagan, C. W. Ickes, and A. S Alving. 1956. Enzymatic deficiency in primaquine-sensitive erythrocytes. *Science* 124:484–489.

Etkin, N. L. 1997. Plants as antimalarial drugs: Relation to G6PD and evolutionary implications. In *Adaptation to malaria: The Interaction of biology and culture*, ed. L. S. Greene and M. E. Danubio, 139–167. Amsterdam, The Netherlands: Gordon and Breach Publishers.

Gray, P. 1986. *Honey from a weed*. New York: Harper and Row.

Haldane, J. B. S. 1938. *Heredity and politics*. New York: W. W. Norton.

———. 1949. Disease and evolution. *La Ricerca Scientifica* 19 (suppl. 1): 3–10.

Katz, S. H. 1987. Fava bean consumption: A case for the co-evolution of genes and culture. In *Food and evolution*, ed. M. Harris and E. B. Ross, 133–159. Philadelphia: Temple University Press.

Katz, S. H., and J. I. Schall. 1986. Favism and malaria: A model of nutrition and biocultural evolution. In *Plants in indigenous medicine and diet:*

*Biobehavioral approaches*, ed. N. L. Etkin, 211–228. Bedford Hills, NY: Redgrave Publishing.

Motulsky, A. G. 1960. Metabolic polymorphisms and the role of infectious diseases in human evolution. *Human Biology* 32:28–62.

Ridley, M. 2000. *Genome: The autobiography of a species in twenty-three chapters*. New York: Harper-Perennial.

Roden, C. 2000. *The new book of Middle Eastern food.* New York: Alfred Knopf.

Salzano, F. M. 1975. *The role of natural selection in human evolution*. New York: American Elsevier Publishing.

Senozan, N. M., and C. A. Thielman. 1991. Glucose 6 phosphate dehydrogenase deficiency: An inherited ailment that affects 1000 million people. *Journal of Chemical Education* 681:7–10.

Sinisalco, M., L. Bernini, B. Latte, and A. Motulsky. 1961. Favism and thalassemia in Sardinia and their relationship to malaria. *Nature* 190: 1179–1180.

Williams, R. J. 1956. *Biochemical individuality: The basis for the genetotrophic concept*. New York: John Wiley and Sons.

Wright, C. A. 1999. *A Mediterranean feast*. New York: William Morrow and Company.

## CHAPTER FOUR

Allbaugh, L. G. 1953. *Crete: A case study of an underdeveloped area*. Princeton: Princeton University Press.

Aravanis, C., R. P. Mensick, A. Corondilas et al. 1988. Risk factors for coronary heart disease in middle-aged men in Crete in 1982. *International Journal of Epidemiology* 17:779–783.

Campos, H., M. D'Agostino, and J. M. Ordovas. 2000. Gene-diet interactions and plasma lipoproteins: Role of apolipoprotein E and habitual fat intake. *Genetic Epidemiology* 20 (1): 117–128.

Kafatos, A., I. Kouroumalis, I. Vlachonikolis, C. Theodorou, and D. Labadarios. 1991. Coronary heart disease risk factor status of the Cretan urban population in the 1980s. *American Journal of Clinical Nutrition* 54:591–598.

Keys, A. B. 1980. *Seven countries: A multivariate analysis of death and coronary heart disease*. Cambridge, MA: Harvard University Press.

Lambraki, M. 2001. *Herbs, greens, fruit: The key to the Mediterranean diet*. Iráklion, Greece: Lambraki.

Psilakis, M., and N. Psilakis. 2000. *Cretan cooking*. Iráklion, Greece: Karmanor.

Psilakis, M., and N. Psilakis. 2000. *Herbs in Cretan cooking*. Iráklion, Greece: Karmanor.

Rackham, O., and J. Moody. 1996. *The making of the Cretan landscape*. New York: Manchester University Press / St. Martins Press.

Renaud, S., M. de Lorgeril, J. Delaye, J. Guidollet, F. Jacquard, N. Mamelle, J.-L. Martin, I. Monjaud, P. Salen, and P. Toubol. 1995. Cretan Mediterranean diet for prevention of coronary heart disease. *American Journal of Clinical Nutrition* 61:1360S–1367S.

Shintani, T. T. 1991. *The Hawaii diet*. New York: Pocket Books / Simon and Schuster.

Stein, G. 1990. *Selected writings of Gertrude Stein*. New York: Vintage.

Trichopoulou, A., A. Kouris-Blazos, T. Vassilakou, C. Gnardellis, E. Polychronopoulos, M. Venizelos, P. Lagiou, M. L. Wahlqvist, and D. Trichopoulos. 1995. Diet and survival of elderly Greeks: A link to the past. *American Journal of Clinical Nutrition* 61:1346S–1350S.

Zamprelas, A., H. Roche, and J. M. E. Knapper. 1999. Differences in postprandial lipaemic response between northern and southern Europeans. *Atherosclerosis* 139 (1): 83–93.

## CHAPTER FIVE

Allison, A. C., and B. Blumberg. 1958. Ability to taste phenylcarbamide among Alaska Eskimos and other populations. *Human Biology* 31:352–357.

Andrews, J. 1984. *Peppers: The domesticated* Capsicums. Austin: University of Texas Press.

Bartoshuk, L. M., V. B. Duffy, and I. J. Miller. 1994. PTC/PROP tasting: Anatomy, psychophysics, and sex effects. *Physiology and Behavior* 56:1165–1171.

Billing, J., and P. W. Sherman. 1998. Antimicrobial functions of spices: Why some like it hot. *Quarterly Review of Biology* 73:3–49.

Chasan, R. 1999. Editorial. *BioScience* 49 (6): 431.

Cotter, D. J. 1982. The scientific contribution of New Mexico to the chile pepper. In *Southwestern agriculture: Pre-Columbian to modern*, ed. H. C. Dethloff and I. M. May Jr., 17–27. College Station: Texas A&M University Press.

Crosby, A. W., Jr. 1986. *Ecological imperialism: The biological expansion of Europe, 900–1900*. Cambridge: Cambridge University Press.

Drenowski, A., and C. L. Rock. 1995. The influence of genetic taste markers on food acceptance. *American Journal of Clinical Nutrition* 62: 506–511.

Esquivel, L. 1989. *Como agua para chocolate*. New York: Doubleday.

Fox, A. L. 1931. Taste-blindness. *Science* 73:14.

Jordt, S.-E., and D. Julius. 2002. Molecular basis for species-specific sensitivity to "hot" chili peppers. *Cell* 108:421–430.

Nabhan, G. P. 1997. While chiles are hot. In *Cultures of habitat: On nature, culture, and story*, 277–284. Washington, DC: Counterpoint Press. Originally published June 1997, *Natural History* 196 (5): 24–27.

Reed, D. R., L. M. Bartoshuk, V. Duffy, S. Marino, and A. Price. 1995. Propylthiouracil tasting: Determination of underlying threshold distributions using maximum likelihood. *Chemical Senses* 20:529–533.

Rozin, P. 1982. Human food selection: The interaction of biology, culture and experience. In *The psychobiology of human food selection*, ed. L. M. Barker, 225–269. Westport CT: AVI Publishing.

Rozin, P., and D. Schiller. 1980. The nature and acquisition of a preference for chile peppers by humans. *Motivation and Emotion* 4:77–101.

Sherman, P. W., and J. Billing. 1999. Darwinian gastronomy: Why we use spices. *BioScience* 49 (6): 453–463.

Snyder, L. H. 1932. The inheritance of taste deficiency in man. *Ohio Journal of Science* 32:436–440.

Tepper, B. J. 1998. Propylthiouracil: A genetic marker for taste, with implications for food preferences and dietary habits. *American Journal of Human Genetics* 63:1271–1276.

Tewksbury, J., and G. P. Nabhan. 2001. Seed dispersal: Directed deterrence by capsaicin in chiles. *Nature* 412:403–404.

Whipple, B., M. Martinez-Gomez, L. Oliva-Zarate, P. Pacheco, and B. R Komisaruk. 1989. Inverse relationship between intensity of vaginal self-stimulation-produced analgesia and level of chronic intake of a dietary source of capsaicin. *Physiology and Behavior* 46 (2): 247–252.

## CHAPTER SIX

Alley, L. 2000. *Lost arts: A celebration of culinary traditions*. Berkeley: Ten Speed Press.

Boushey, C. J., S. A. A. Beresford, G. S. Omenn, and A. G. Motulsky. 1995. A quantitative assessment of plasma homocystine as a risk factor for cardiovascular disease. *Journal of the American Medical Association* 274 (13): 1049–1067.

Carper, J. 1988. *The food pharmacy*. New York: Bantam Books.

Choumenkvitch, S. F., J. Selhub, P. W. F. Wilson, J. I. Rader, I. Rosenberg, and P. F. Jacques. 2002. Folic acid intake from fortification in United States exceeds predictions. *Journal of Nutrition* 132:2792–2798.

Cok, I., N. A. Kocabas, S. Cholerton, A. E. Karakaya, and S. Sardas. 2001. Determination of coumarin metabolism in Turkish population. *Human and Environmental Toxicology* 20:179–184.

Etkin, N. L. 1986. Multidisciplinary perspectives in the interpretation of plants used in indigenous medicine and diet. *Plants in indigenous medicine and diet: Biobehavioral approaches*, ed. N. L. Etkin, 2–28. Bedford Hills, NY: Redgrave Publishing.

Grierson, B. 2003. What your genes want you to eat. *New York Times Magazine*, May 4, 77–79.

Johns, T. 1990. *With bitter herbs they shall eat it: Chemical ecology and the origins of human diet and medicine*. Tucson: University of Arizona Press.

McKibben, B. 2003. *Enough: Staying human in an engineered age*. New York: Times Books / Henry Holt.

McKusick, V. A. Accessed 2002. Online Mendelian Inheritance in Man database, entries for warfarin resistance, albumin, and coumarin 7-hydroxylase. www.ncbi.nlm.nih.gov/entrez (site now discontinued).

Mitchell, M. 1998. *Diné biké yyahdóó ch'il nanisé altaas'éí: Plants of Navajoland*. Chinle, AZ: Chinle Curriculum Center.

Motulsky, A. G. 1996. Nutritional ecogenetics: homocystine-related arteriosclerotic vascular disease, neural tube defects, and folic acid. *American Journal of Human Genetics* 58:17–20.

Nallamouthou, B. K., A. M. Frederick, M. Rubenfire, S. Saint, R. R. Bandekar, and G. S. Omenn. 2000. Potential clinical and economic effects of homocystine lowering. *Archives of Internal Medicine* 160:3406–3412.

Olson, S. 2002. *Mapping human history: Discovering the past through our genes*. Boston: Houghton-Mifflin.

Raichelson, R. M. 1986. Coumarin-containing plants and serum albumin polymorphisms: Biomedical implications for Native Americans of the Southwest. *Plants in indigenous medicine and diet: Biobehavioral approaches*, ed. N. L. Etkin, 229–241. Bedford Hills, NY: Redgrave Publishing.

Starvic, B. 1997. Chemopreventive agents in foods. *Functionality of food phytochemicals*, ed. T. Johns and J. T. Romero, 53–88. New York: Plenum Press.

Weil, A. 2000. *Eating well for optimum health*. New York: Alfred Knopf.

## CHAPTER SEVEN

Brand-Miller, J. C., J. Snow, G. P. Nabhan, and A. S. Truswell. 1990. Plasma glucose and insulin responses to traditional Pima Indian meals. *American Journal of Clinical Nutrition* 51:416–420.

Brand-Miller, J. C., and A. W. Thorburn. 1987. Traditional foods of Australian aborigines and Pacific Islanders. In *Nutrition and health in the tropics*, ed. C. Rae and J. Green, 262–270. Canberra, Australia: Menzies Symposium.

Brand-Miller, J. C., and S. Colaguiri. 1994. The carnivore connection: Dietary carbohydrate and the evolution of NIDDM. *Diabetologica* 37:1280–1286.

———. 1999. Evolutionary aspects of diet and insulin resistance. In *Evolutionary aspects of nutrition and health: diet, exercise, genetics, and chronic disease*. Vol. 84 of *World review of nutrition and diet*. Basel, Switzerland: Karger.

Cordain, L., J. C. Brand-Miller, and N. Mann. 2000. Scant evidence of periodic starvation among hunter-gatherers. *Diabetologica* 24 (3): 2400–2408.

Cowen, R. 1991. Desert foods offer protection from diabetes. *Science News* 32:12–14.

Diamond, J. 1992. Sweet death. *Natural History* 2:2–7.

Gladwell, M. 1998. The Pima paradox. *New Yorker*, February 2, 41–53.

Infante, E., A. Olivo, C. Alaez, F. Williams, D. Middleton, G. de la Rosa, M. J. Pujo, C. Duran, J. L. Navarro, and C. Gorodezky. 1999. Molecular analysis of HLA class I alleles in Mexican Seri Indians: Implications for their origin. *Tissue Antigens* 54:35–42.

Nabhan, G. P. 2004. *Cross-pollinations: The marriage of science and poetry*. Minneapolis: Milkweed Editions.

Nabhan, G. P. , C. W. Weber, and J. Berry. 1979. Legumes in the Papago-Pima Indian diet and ecological niche. *Kiva* 44173–178.

Neel, J. V. 1962. Diabetes mellitus: A 'thrifty genotype' rendered detrimental by progress. *American Journal of Human Genetics* 14 (4): 353–362.

———. 1998. The 'thrifty genotype' in 1998. *Perspectives in Biology and Medicine* 42:44–74.

O'Dea, K. 1984. Marked improvement in carbohydrate metabolism in diabetic Australian aborigines after temporary reversion to traditional lifestyle. *Diabetes* 33:596–603.

Seppa, N. 2002. Gene tied to heightened diabetes risk. *Science News* 158:212.

Swinburn, B. A., V. L. Boyce, R. N. Bergman, B. V. Howard, and C. Borgardus. 1993. Deterioration in carbohydrate metabolism and lipoprotein changes induced by a modern, high fat diet in Pima Indians and Caucasians. *Journal of Clinical Endrocrinology and Metabolism* 73 (1): 156–164.

Villela, G. J., and L. A. Palinkas. 2000. Sociocultural change and health status among the Seri Indians of Sonora, Mexico. *Medical Anthropology* 19:147–172.

Weber, C. W., R. B. Arrifin, G. P. Nabhan, A. Idouraine, and E. A. Kohlhepp. 1996. Composition of Sonoran Desert foods used by Tohono O'odham and Pima Indians. *Journal of the Ecology of Food and Nutrition* 26:63–66.

White, M. 2000. New insights into type 2 diabetes. *Science* 289:37–39.

## CHAPTER EIGHT

Enos, E., S., K, Johnson, and S. Enos. 1995. *A handbook of kalo basics*. Waianae, Hawaii: Taro Top Publications.

Greer, N. 2002. Kalo farming: Lessons in cultural survival, wetlands management and traditional ecological knowledge. Wetlands and Man in Hawaii workshop, October 31–November 1, Wildlife Society, Honolulu.

Jackson, F. L. C. 2001. The Human Genome Project and the African-American community: Race, diversity and American science. In *The Human Genome Project and minority communities*, ed. R. A. Zinkas and P. J. Balint, 35–52. Westport, CT: Praeger.

Krauss, B. H. 1993. *Plants in Hawaiian culture*. Honolulu: University of Hawaii Press.

McKibben, B. 2003. *Enough: Staying human in an engineered age*. New York: Times Books / Henry Holt.

Murray, R. F., Jr. 2001. Social and medical implications. In *The Human Genome Project and minority communities*, ed. R. A. Zinkas and P. J. Balint, 53–75. Westport, CT: Praeger.

Olson, S. 2002. *Mapping human history: Discovering the past through our genes*. Boston: Houghton-Mifflin.

Shintani, T. T., S. Beckman, A. C. Brown, and H. K. O'Connor. 2001. The Hawaii Diet. *Hawaii Medical Journal* 60:69–73.

Shintani, T. T., S. Beckman, J. Tang, H. K. O'Connor, and C. K. Hughes. 1999. Waianae Diet program: Long-term follow-up. *Hawaii Medical Journal* 58:117–121.

Shintani, T. T., C. K. Hughes, S. Beckman, and H. K. O'Connor. 1991. Obesity and cardiovascular risk intervention through the *ad libitum* feeding of traditional Hawaiian diet. *American Journal of Clinical Nutrition* 53:1647S-1651S.

Weissman, G. 2002. *The year of the genome: A diary of the biological revolution*. New York: Times Books / Henry Holt.

# INDEX

# About Island Press

Since 1984, the nonprofit Island Press has been stimulating, shaping, and communicating the ideas that are essential for solving environmental problems worldwide. With more than 800 titles in print and some 40 new releases each year, we are the nation's leading publisher on environmental issues. We identify innovative thinkers and emerging trends in the environmental field. We work with world-renowned experts and authors to develop cross-disciplinary solutions to environmental challenges.

Island Press designs and implements coordinated book publication campaigns in order to communicate our critical messages in print, in person, and online using the latest technologies, programs, and the media. Our goal: to reach targeted audiences—scientists, policymakers, environmental advocates, the media, and concerned citizens—who can and will take action to protect the plants and animals that enrich our world, the ecosystems we need to survive, the water we drink, and the air we breathe.

Island Press gratefully acknowledges the support of its work by the Agua Fund, Inc., The Margaret A. Cargill Foundation, Betsy and Jesse Fink Foundation, The William and Flora Hewlett Foundation, The Kresge Foundation, The Forrest and Frances Lattner Foundation, The Andrew W. Mellon Foundation, The Curtis and Edith Munson Foundation, The Overbrook Foundation, The David and Lucile Packard Foundation, The Summit Foundation, Trust for Architectural Easements, The Winslow Foundation, and other generous donors.